Nicholas II

NICHOLAS II
Last of the Tsars

Marc Ferro

Translated by Brian Pearce

New York Oxford
OXFORD UNIVERSITY PRESS
1993

Oxford University Press

Oxford New York Toronto
Delhi Bombay Calcutta Madras Karachi
Kuala Lumpur Singapore Hong Kong Tokyo
Nairobi Dar es Salaam Cape Town
Melbourne Auckland Madrid

and associated companies in
Berlin Ibadan

Library of Congress Cataloging-in-Publication Data
Ferro , Marc.
Nicholas II : the last of the tsars / Marc Ferro ;
translated by Brian Pearce.
p. cm.
Includes bibliographical references and index.
ISBN 0-19-508192-7
1. Nicholas II, Emperor of Russia, 1868-1918.
2. Russia—Kings and rulers—Biography.
3. Russia—History—Nicholas II, 1894-1917.
I. Title. DK258.F47 1993
947.08′092—dc 20   92-41440

2 4 6 8 9 7 5 3 1
Printed in the United States of America
on acid-free paper

To Martine Godet
and Annie Goldmann

Contents

Illustrations

PHOTOGRAPHS

Alexander III and his family.
Nicholas II and the Prince of Wales in 1909.
Nicholas II, a Maharaja and Prince George of Greece.
Empress Alexandra prepares for her first ball.
Nicholas II with the Tsarevich and the Grand Duchesses.
The coronation.
Entry by Nicholas II into Balmoral.
Their Majesties leaving the Trinity Cathedral.
Members of the government and of the zemstvos.
Stolypin.
Plehve.
Witte. (*Roger-Viollet*)
Sipyagin.
Nicholas II with his officers.
Opening of the First Duma.
Execution of revolutionaries.
Leaflet: 'The Bloody Tsar'.
Rasputin.
Rasputin with his 'court'.
Official photograph of the imperial children.
The Tsar's daughters as prisoners at Tsarskoe Selo.
At the Ipatiev house.
Maria, 1913.

Forty years later?
Anastasia. (*Roger-Viollet*)
Anna Anderson, who claimed to be Anastasia.
The Czechoslovaks and Whites take Ekaterinburg from the Reds.

All photographs with the exception of those of Witte and the Grand-Duchess Anastasia are reproduced by kind permission of Éditions Payot.

Acknowledgements

This work would not have been written without the encouragement I received, first and foremost, from the little seminar to which I presented my first conclusions – S. C. Ingerflom, T. Kondratieva, V. Garros, M. Ferretti, A. Salomoni, J. Scherrer, M. H. Mandrillon and A. Berelovitch. The development and construction of the work were closely followed also by Martine Godet and Daniel Milo. Vonnie read it and made corrections, as usual, and C. Murco deciphered and revised it.

Hélène Kaplan and her colleagues at the BDIC kept me informed, of course, about the latest publications. J. Grivaud and J. Catteau drew my attention to a number of useful books and articles, for which I thank them. My gratitude is also due to Claude Durand and Marcel Laignoux, and to Alexis de Durazzo, Prince of Anjou, who allowed me to see some of the documents in his possession and answered my questions.

At Éditions Payot, Dominique Missika showed me what is meant by a good publisher. Also to be thanked is a certain anonymous reader who, being forthright with her criticisms, led me to reconsider an entire chapter. I am grateful to them and to Catherine Ritchie, who revised the entire manuscript, bringing this ordeal to its conclusion.

Finally, I wish to thank Brian Pearce, who translated the English edition, for having read the original text critically and for correcting a number of mistakes.

FAMILY TREE OF THE ROMANOVS

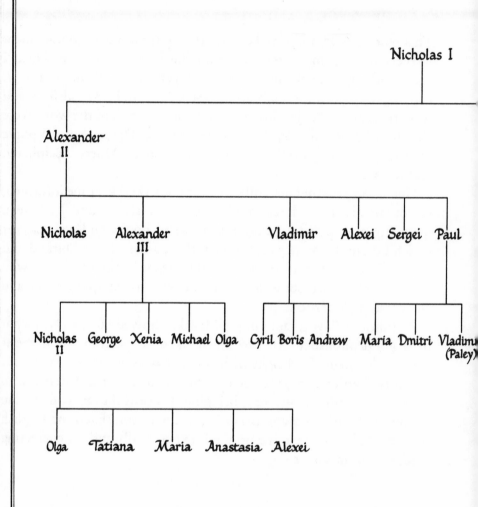

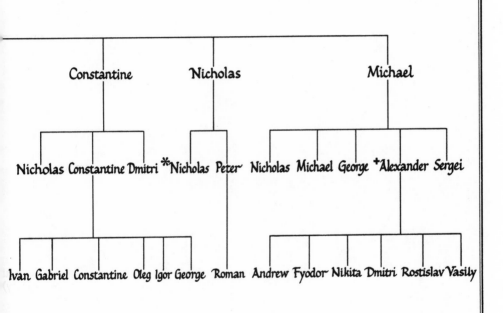

Constantine   Nicholas   Michael

Nicholas   Constantine   Dmitri   *Nicholas   Peter   Nicholas   Michael   George   +Alexander   Sergei

Ivan   Gabriel   Constantine   Oleg   Igor   George   Roman   Andrew   Fyodor   Nikita   Dmitri   Rostislav   Vasily

*Women members of the family are shown
only if mentioned in the book.*

* Nicholas Nikolaevich ('Nikolasha')
\+ Alexander Mikhailovich ('Sandro')

Introduction

When, in 1894, he learned that he was to ascend the throne, Nicholas burst into tears. What he feared most had come upon him, by God's will. 'Oh, Sandro,' he said to Alexander, his cousin and boyhood friend, 'I am not ready to be Tsar,' and he continued to sob bitterly.

His dream had been to be a sailor, to travel, to go on voyages round the world. Instead, he would have to attend Councils of Ministers, read reports, decide, govern, act.

As heir to the throne, Tsarevich, he liked ceremonies and festivals, first nights at the opera, the life of high society generally. In that setting he was at ease, faultless in his conduct, discreet, jovial, charming. But he avoided all serious talk, and especially talk about the situation in the country: ministers were appointed to see to all that. His only duty was to keep Russia great, and to maintain intact the powers that God had entrusted to him. Besides, in the family, at the table where his father Alexander III presided, these matters were never raised, and there was certainly no talk of politics.

Yet in Russian society, where, after the reforms of Alexander II, people had looked forward to some sort of springtide, and had been deeply disappointed by the uncompromising reaction that followed his assassination in 1881, a demand for politics was fermenting that would not be denied.

For decades an infinite number of projects and programmes and works of literature, every one of which constituted a sign, an indictment, had shown that this was so. Bakunin, Belinsky, Tolstoy, Mikhailovsky, Chernyshevsky, Dostoevsky, Plekhanov were the master-thinkers of this movement, along with the artists of their time. It was their ideas that had nourished the generation to which belonged Nicholas Alexandrovich, the Tsarevich, born in 1868, who was thirteen when his father became Tsar and twenty-six when he died. In Europe, his cousin Wilhelm II was nine years older, and the Duke of York, the future George V, three years younger.

The new Tsar was not interested in this ferment of ideas that was stirring the country. A kind of basic incomprehension stood between him and cultivated society, whose aspirations, he felt, did not have as their principal objective the preservation of his powers.

This reign which had begun as an obligation imposed by God thus became a nightmare. Nicholas had to face two revolutions, to endure all round his person dozens of assassinations and attempted assassinations, to preside over an assembly, the Duma, that he had not wanted to call into existence, and to participate in its sessions or in those of endless Councils of Ministers. He also had to go to war twice, whereas he would have liked to figure as an apostle of peace. After a lengthy captivity he died, murdered. Both before and after his abdication, his chief care had been the health of his son and sole heir, an incurable haemophiliac.

He seemed indifferent, concerned most of the time with such questions as what uniforms he should wear some evening, or which ballerina was going to dance in *Swan Lake*, or who would accompany him to the hunt next day. He has been seen as lying under the spell of Rasputin, aloof from the world's fate, weak and irresolute, characterless. In a famous passage, Trotsky compared him to Louis XVI, 'though less intelligent'.

This picture is only half correct. Nicholas II was perfectly aware of what he was doing when he initiated the Hague Peace Conference, or when war broke out with Japan. He was perfectly aware, too, after 'Bloody Sunday' (his soldiers had fired on a procession of people who were coming to submit a petition to him), when he 'forgave his people for having rebelled against him'. Aware, likewise, in 1914, when war began despite his efforts, and no less so when he ordered repressive measures to be taken in February 1917 as in 1905. After his abdication he expressed regret that the Provisional Government, after reintroducing the death penalty, had nevertheless failed to apply it adequately.

From 1894 until his end his behaviour showed him to be conformist and traditionalist. In everything he did and in all his comments and confidences we observe the same taste for the order, ritual and ceremony identified with the intangible grandeur of autocracy. He hated everything that might shake that autocracy: the intelligentsia, modernity, the Jews, the religious sects. Towards those who did not conform with his ideas he hardened his heart, whereas he looked with tender kindness on those he loved.

Dominated by his wife, like Louis XVI, this ruler who came to be called Nicholas 'the Bloody' was not bloodthirsty. He was quick to deplore the cruel consequences of measures he decreed. He considered that he had simply done his duty, shooting people as a matter of scruple.

Above all, he was always at the receiving end of events, never their initiator. History never stopped hustling him. He believed that he had a duty to oppose change. He wished that people would cease importuning him with all those projects for a constitution and would let him live quietly with his family.

In fact, on that day in 1917 when two deputies from the Duma came to ask him to abdicate in order to save the dynasty, after having refused so often, either gently or stubbornly, to modify the way the regime operated, he now signed away his throne without arguing.

On the platform of Pskov railway station, the officers of his personal guard, who had been told what had happened, choked back their tears as Nicholas II returned their salute and stepped briskly into the train, saying: 'At last I shall be able to go and live at Livadia.' A witness even heard him whistle softly, as he had done when told of the murder of Rasputin. The Empress had collapsed in tears, but Nicholas went off cheerfully. Everyone agreed in saying that the Tsar knew how to hide his feelings.

'I order that a stop be put to these disorders, which are intolerable in wartime,' he had telegraphed when he learned of the revolutionary events in Petrograd on 25 February 1917. Then, after taking lunch, he had written a letter to his wife Alexandra,[1] to tell her the state of

---

1. Alix of Hesse-Darmstadt, Nicholas's betrothed, was converted to the Orthodox faith on 22 October 1894 and became Grand Duchess Alexandra Fyodorovna. She married the Tsar on 14 November 1894, Old Style (26 November, New Style).

his health; and, after that, he had ordered the train to be halted so that he might go for a short walk in the woods and pick mushrooms.

And yet some of his writings give proof that he was able to see, understand and judge. On 2 March 1917 he noted in his private diary: 'I left Pskov at one o'clock, my heart heavy from what has happened. All around me are treason, cowardice and deceit.'

An enigmatic figure, Nicholas II was one of those individuals who are burdened with a destiny that they have taken upon themselves as a duty decreed once for all time, and that separates them from a world undergoing change before their eyes. Nicholas II was not blind to this, but he considered that his duty consisted entirely in showing respect for the past and humility before God by yielding up none of his powers – unless constrained and forced to do so, as in 1905. It would be better to abdicate than to betray his duty.

History knows personages like this who, in the name of their faith, remain alien to all the warnings history gives them. For Nicholas II and his circle the enemy meant, first and foremost, the terrorists and other nihilists, together with all who sympathized with or supported them: in short, all who shared, to one extent or another, the new ideas, and who advocated change. The Tsar made no distinction between the socialists who were remote from him and the liberals who were closer. He refused to listen to advisers or ministers who proposed that a constitution be introduced in order to neutralize them.

When, in 1905, following the advice of his Prime Minister, Witte, he had 'granted' a Duma, so as to calm agitated spirits, and yet, despite this concession by him, strikes redoubled, he exclaimed: 'How could a man who is said to be so intelligent have made such a mistake?' And when Stolypin, his other great minister, though well known for his firmness, urged him to enter into a dialogue with the Duma, he refused. Later, he was to tell his cousin Sandro: 'I believe in no one but my wife.' And his wife told him again and again that, as an autocrat, he was by definition debarred from sharing his power.

*

An ordinary man, Nicholas II carried a unique burden, which crushed him. The enigma of his death is not unconnected with this fact.

Along with their bodies the wheels that had been used to execute them and the very sleighs on which the condemned men had been drawn to the place of execution were burned ... Nothing of him must be left ... Already the house where he lived had been reduced to ashes. His name vanished ... The nearby river was given a new name ... And, so as to obliterate all memory of the executions, great celebrations were later held in those same places, to express joy in the peace that had been established.

This passage does not relate to the death of Nicholas II in 1918 but to that of Pugachev, executed in 1775. It has strange resonances, for, over 140 years later, at Ekaterinburg, things happened in much the same way.

No trace was left of the dead. Was this a way of denying that they existed? Of preventing their return from the dead?

But did they really die, or are they *all* dead?

At this crossroads, where legend intersects with news items, and where the Bolshevik regime re-enacts the deeds of Holy Russia, doubt, as will be seen, declines to go away.

1. Society versus the Autocracy: Nicholas 'the Unlucky'

Young 'Nicky' was thirteen when he saw his grandfather Alexander II, his legs smashed by a nihilist's bomb, dying before his eyes. 'The Emperor lay on the couch ... Three doctors were fussing around, but science was obviously helpless ... He presented a terrible sight, his right leg torn off, his left leg shattered, innumerable wounds all over his head and face ... One eye was shut, the other expressionless. I clung to Nicky's arm,' his cousin Grand Duke Alexander tells us. 'He was deathly pale in his blue sailor's suit. His mother, stunned by the catastrophe, was still holding a pair of skates in her trembling hands ... That expressionless eye was still staring fixedly.'

Such was Nicholas's first encounter with history.

On that day, 1 March 1881, all the family were there around Alexander II: his son Alexander Alexandrovich, who was to be Alexander III, and his grandson Nicholas Alexandrovich, the future Nicholas II.

When Alexander II ascended the throne, his father Nicholas I had said to him before he died: 'I am not handing over command of the empire to you in the favourable circumstances I should have wished. I am bequeathing to you much trouble and care.'

Nicholas I's own reign (1825–55) had opened with a sort of failed revolution, the revolt of the Decembrists, young officers of the noble class who were influenced by the ideas of the French Revolution, and it had ended with a setback, defeat in the Crimean War, which everyone, the Tsar first and foremost, blamed on an absolutist system that, though giving an illusion of strength, had revealed its limitations. As 'gendarme of Europe' for a quarter of a century, the Tsar had embodied the spirit of the Holy Alliance more than any other ruler. His exploits had included the crushing of the Polish insurrection in 1831 (subject of the famous phrase: 'Order reigns in

Warsaw') and the crushing of the Hungarian insurrection of 1849, a 'service' rendered to his cousin, the Emperor of Austria.

In internal affairs, having grasped that the ideas of the Decembrists, drawn from the arsenal of the Enlightenment, had been spread as a result of Napoleon's conquests, Nicholas I had sought to lock the gates of his empire as firmly as he could. He earned the nickname of 'Don Quixote of Absolutism'. The effect of his stern laws was to precipitate the growth of a revolutionary movement.

And, whereas elsewhere in Europe the revolutionary spirit was merely republican or democratic, in Russia it became nihilist and terrorist.

This extremist tendency developed with all the greater force because the reforms introduced by Nicholas I affected, essentially, only the Crown estates, on which he took steps to improve the lot of the peasants. The great mass of the peasantry, those who were not settled on the Crown lands, felt even more bitterly than before the condition in which they had been left. A caricature of the time shows some *pomeshchiki* (landlords) playing a game of cards. The stakes are bundles of serfs, tied up like sticks of asparagus. Gogol described in *Dead Souls* the lot of these serfs, subject to a bureaucracy such that, in the interval before their deaths were registered, shady serf-owners sold them to speculators, who resold them in their turn.

This muzhik, 'poorer than the Bedouin, poorer than the Hebrew', to use Herzen's words, 'had nothing with which he could console himself'. There, according to the father of Russian socialism 'lay the secret of his potential revolutionariness'. He became the stake in the game played between Tsardom and the revolutionaries.

Alexander II, who succeeded Nicholas I and reigned from 1855 to 1881, was well aware of this fact, and managed to make the *pomeshchiki* realize that it was 'to their interest that reform should come from above rather than from below'. By his manifesto of 1861 he abolished serfdom, not only out of humanitarianism but also in order to prevent another peasant revolt like Pugachev's. The peasants could become *de facto* masters of the land they tilled in return for a staggered redemption payment made to the state, which would indemnify the landowners directly. For the peasants, this payment proved to be a burden they were unable to bear. Nevertheless, Alexander II was 'the Tsar Liberator'.

He also made some concessions to the cultivated strata of society, unmuzzling the press to a certain extent, developing education and granting a modicum of autonomy to the local representative assemblies called *zemstva*, in Russia proper, on the express condition that they concerned themselves with local problems only. As compared with the past, that meant a great deal.

Compared, however, with the expectations of cultivated society – nobility, bureaucracy, intelligentsia – it was nothing. These educated people were able to compare their fate with that of people in the West, where, despite the reactions of 1815–30 and 1850–70, the citizens of France or Britain, of Prussia or Piedmont, possessed parliaments and enjoyed political liberties. To be sure, Alexander II's work might have culminated in a form of popular representation not restricted merely to the management of local affairs. But this did not happen, because the opposition of a section of the landed aristocracy, hostile to reform, the fresh insurrection of the Poles in 1863, the resistance of certain ministers and the Tsar's own hesitation paralysed such an evolution.

At the same time the cultivated classes were growing impatient. Turgenev described in *Fathers and Sons* the unbearable waiting endured by the young generation, prevented from influencing their country's future and disposed to blame their parents for having submitted meekly to this situation. Dostoevsky, who had been a convict in Siberia, also analysed the demands of this generation of nihilists who challenged everything – state, family, morals – through his depiction of Stavrogin, a caricature 'truer than the reality'. Pisarev, Bakunin and others inspired an extremist revolutionary movement that, with Tkachev, would use terrorism to overthrow the hated regime.

It is true that Alexander II was 'helped' by the police, the notorious Third Section of His Majesty's Chancellery. This organ ferreted everywhere, carrying out the orders of the head of the gendarmerie, opening letters, infiltrating its agents even into boudoirs, making itself hated and turning nihilists into terrorists. Its omnipotence enabled it to function as a redoubtable engine that mowed down human lives with a circumspection such as history has rarely seen.

Turned over subsequently to the Ministry of the Interior and renamed the Okhrana, it was so effective as to make Russia the first police state in the West.[1] It controlled a numerous body of gendarmes, assisted by a multitude of secret agents, both permanent and temporary. Spies in military uniform kept an eye on the army officers. All houses in towns had a porter, and all these porters were linked with the police, and made a daily report to them. No one could be sure of the discretion even of his best friend. The correspondence of the Empress herself was not exempt from the attentions of the *cabinet noir*. From the Niemen to the Pacific the close-meshed network of a thorough tyranny held the population in its grip. Nearly a century of this system implanted in the Russians a feeling of suspicion, mistrust and reticence.

Last winter, at Monaco, than which no place in Europe is visited by larger numbers of Russians, I was talking with a landholder of the Don about his country. We were alone. Suddenly, a stranger came towards us. My friend immediately changed the subject, began to speak of theatres, concerts: in the newcomer, from his features and general look, he thought he recognized a countryman.

This scene, reported by Anatole Leroy-Beaulieu in a work that is undoubtedly the best depiction of the Russian empire in this period, dates not from the Russia of Stalin or Lenin but from the late nineteenth century.

Just as in France the excesses of the *ancien régime* gave rise to the spirit of revolution, in Russia it was the police who taught people to be nihilists. So hated were they that when, in 1878, Vera Zasulich shot at General Trepov, who had had a political prisoner flogged, a court acquitted the young woman amid public acclamation. A boundary had been crossed; society and the regime had split apart.

1. Russia was the first police state in the West, but not the first in history since the India of the Mauryas and the Guptas, in the fourth century BC, had been such a state.

Thus encouraged, so to speak, nihilists and terrorists now made the Tsar their target, and after several attempts at last struck down Alexander II. The assassination of the Tsar in 1881 marked with a deep wound the 'sacred' bond linking Tsardom with cultivated society. Sons of that society had dared to take the life of a Tsar, and not the tyrant Tsar Nicholas I, either, but the 'Tsar Liberator'. It was proof that the point of no return had been reached.

Alexander III, the murdered Tsar's son, would have liked to be a tolerant and reforming ruler. But his father's fate altered his views and for thirteen years (1881–94) he governed Russia with the Cossack whip, the *nagaika*. True, his tutor (who was also to become the tutor of his son, the future Nicholas II) Constantine Pobedonostsev, the Procurator-General of the Holy Synod, made an important contribution to this reversal of attitudes, which had been developing for several years, so that the bomb of 1881 served as pretext rather than cause for the change.

The ideas that came from Europe were not suitable for Russia, the Tsar's ministers explained. The Tsar should rule for the people, along with his peasants. He should revive the myth of the Little Father Tsar, the Protector, who had just been murdered 'by the nihilists and the Jewish plague'. Were the pogroms that followed the assassination in 1881 fomented by the Sviashchennaia Druzhina (Sacred Company), which was created in 1881 and was closely linked with the Okhrana? This body was dissolved, however, not on account of its excesses but because, for fundamentalists of the autocracy, the Tsar had no need of any special, secret guard to protect him by answering murders with other murders. The state itself should suffice for that purpose. Alexander III should present himself as a Russian Tsar, not as a European emperor, and this was a role that fitted him like a glove.

With these reforms and resistances to reforms, these plans either carried out or abandoned, the impression was created, however, in 1881 that a fresh 1789 was in the offing. The revolutionary circles were in ferment. If there should be a popular uprising, ought they to go along with the masses, as the Narodniks urged, or ought they to forestall the masses, act on their behalf so as to avoid the false steps taken by the French Revolution, meaning the Terror, Thermidor and eventually Bonaparte? And advocates and opponents of 'individual terror' argued about the best way to overthrow the regime.

Well, since the nihilists wanted to terrorize the Romanovs, Alexander III decided that the Romanovs would terrorize the nihilists. In 1881, before his assassination, Alexander II had been working with his ministers Abaza and Loris-Melikov on a project for a constitution such as the cultivated classes wished to see. But Pobedonostsev succeeded, by patient persistence, in turning Alexander III away from that idea. He effected a sort of *coup d'état* in miniature by completely reversing the position of the new Tsar, so that the two ministers of Alexander II who in the morning thought they were already 'constitutional' ministers found in the evening of that day that they were still the Tsar's ministers in the time-honoured sense.

Alexander III governed Russia by means of terror. He succeeded fairly well in this, thanks to the ability of the head of the police department, Plehve, who consecrated the system of arbitrary rule by the Okhrana. From 1883 the country was in a 'state of reinforced security', under which individual liberties could be suspended, civil cases transferred to military tribunals, citizens assigned to compulsory places of residence or sent into exile by mere administrative decision, and any publication considered dangerous could be banned. This 'temporary law' was renewed year after year. When Nicholas II became Tsar in 1894, 5,400 persons had been exiled or sentenced to hard labour. Young women were kept under particular surveillance, since they accounted for 158 of the persons charged, up to that date, with attempts on the life of the Tsar – one-quarter of the total (Pahlen Report).

This policing of people's thoughts was accomplished by a Great Russian and Orthodox reaction. In the universities of Poland and the Baltic provinces use of the Russian language was made obligatory, and corresponding measures were initiated in Finland. Inspired by Pobedonostsev, the Orthodox revival found expression in official anti-Semitism: a *numerus clausus* was instituted in 1887 in order to restrict the number of Jews in the universities. 'Let us never forget that it was the Jews who crucified Jesus,' observed Alexander III, writing these words at the foot of the decree.

Catholicism in Poland and Protestantism in the Baltic provinces were also subjected to surveillance, these two religions being suspect, as was shown by the expression used to designate them: 'foreign denominations' (*inostranniia ispovedeniia*). Though the different forms of Christianity were allowed freedom of worship in Russia, proselytism was the exclusive privilege of the Orthodox Church. Under Alexander III the Evangelical Alliance complained about this and asked that all the Christian churches be allowed full freedom. Pobedonostsev replied that, in fact, such freedom prevailed, 'except for the right to engage in unrestricted propaganda'.

Russia having received her vital principle from the Orthodox faith, it is her sacred duty ... to shield the church ... The Western denominations, far from laying aside their domineering ways, are always ready to attack not only our country's power but her unity ... Russia will never allow that the Orthodox Church's children should be taken from her to be enrolled in alien flocks.

In the Baltic provinces the policy was rather to secularize, so as the more easily to Russify later. This haughty watchfulness was, of course, also applied to the Uniate Church in the Ukraine and to the Armenian Church, and was resented as a blow aimed at the national identities of the peoples concerned.

The religious reaction struck likewise at schismatics, such as the Old Believers, and sects, like the Dukhobors – a rationalistic sect who condemned the excessive ritualism of most Orthodox ceremonies and sacraments. They called themselves spiritual Christians, were distinguished by their contempt for the traditional forms of worship and did not recognize any priesthood. 'The schismatics would go to the block for the sign of the cross with two fingers. As for us, we don't cross ourselves at all, either with two fingers or with three, but we strive to gain a better knowledge of God.' The Dukhobors were said to hold that governments were made for bad people only, whereas *they* were obliged to obey only the eternal law that God had inscribed on the tablets of their hearts. They advocated offering non-violent resistance to injustice.

These simple notions we find, too, in the writings of Leo Tolstoy, who also tried to find the truth in the Gospels and wanted to preach the looked-for revolution, but by means of words of love. He appealed to the new Tsar, Alexander III, to spare his father's murderers. 'To combat them you need to meet them on the field of ideas. Their ideal is universal well-being, equality, liberty. To combat them you must put forward another ideal, higher than theirs, greater and more generous.'

But he was not listened to. He was not even heard; for the Tsar – Alexander III no less than Nicholas II later – was surrounded by a circle which, being made up of his family, the court and the administration, cut him off from the rest of the world. In consequence of this the Tsar's omnipotence was to some degree fictitious. Prince Trubetskoy wrote *c.* 1900:

There is the autocracy of the police, that of the provincial governors, and that of the ministers. The Tsar's own autocracy is non-existent, because he knows only what a complicated system of filters allows him to know. Therefore this Tsar-autocrat is more restricted in his real power, owing to his actual ignorance of the situation in the country, than a monarch who is directly in touch with the elected representatives of his nation.

By its ultra-conservatism the autocracy thus gave rise both to terrorism and to non-violent resistance. In Alexander III's reign the political police, the Okhrana, made no distinction between these. Adherents of both tendencies were condemned, excommunicated or sent to forced labour in Siberia. From *this* angle the power of the autocracy seemed to be unlimited.

Under the influence of his mentor Pobedonostsev, Alexander III showed himself sensitive to anything that might weaken the autocratic principle. The enemy was marked down: namely, all those publicists and other writers who, infected with liberal ideas brought in from western Europe, had adopted extreme forms of these ideas – nihilism, socialism – combined with practices *à la russe* – terrorism.

These ideas were even penetrating the bureaucracy, that other face of the nobility, as the reformer I. Samarin called it: 'The bureaucrat is always just a nobleman in uniform, and the nobleman a bureaucrat in a dressing-gown.' This bureaucracy had been infected ever since Alexander II had introduced his 'reforms'.

It was 1881, the very year in which that ruler had been murdered, that saw the publication of the dictionary compiled by V. Dahl, a learned lexicographer without any special commitment, who introduced the word 'liberal' in his work. He defined the liberal as 'a free-thinker in politics: someone who desires much liberty for the people, with a form of self-government'. This was the very definition of subversion so far as Alexander III and the theoreticians of threatened autocracy were concerned – Pobedonostsev, Count D. Tolstoy, N. Katkov, V. Meshchersky. Through the working of the reforms these liberal ideas had made their way into the bureaucratic apparatus, especially the State Council, the 'nursery' from which it was recruited, and the members of which were chosen by the sovereign. In the hierarchy of institutions this council was the body through which passed proposals for laws that were to be submitted to the Tsar and those that he approved. It was a kind of forum wherein, of necessity, all the state's problems were discussed, and discussed by men who knew what they were talking about; 46 per cent of the councillors had had higher education.

Alexander III considered that the entire system had become cank-ered since the instances closest to him had started to discuss and dispute about everything. 'We ought to have one single newspaper which would say what should be thought,' joked Kozma Prutkov, a character invented by A. K. Tolstoy. In fact, the regime was entrapped by its own definition of autocracy. Essentially, the auto-cratic sovereign has no master but himself, appointed by God. As head of the church he is not accountable to it, unlike a Catholic sovereign, nor does he owe anything to his nobles. The state is, in a sense, his personal property, the proper functioning of which he ensures, for the good of the Russian people. If he goes beyond that, he is a despot, like Tsar Paul. Should he restrict his own power? Even if he were to consider doing so, such restriction could not be imposed upon him from without. But, in theory, he could do it, as in fact Alexander II almost did.

For his part, Nicholas II held scrupulously to these principles, and he never took the initiative in restricting his power. Though he did restrict it under pressure in 1905, in 1917 he preferred to abdicate rather than yield any further.

For young Nicholas, Alexander III, that rough and moody father, was tenderness personified. Whereas his mother maintained with her son only such relations as propriety dictates, the 'bear' Alexander secretly went up to Nicholas's bedroom to make a fuss of him. And young Nicholas adored this 'bogyman' who rose at seven, washed in ice-cold water, drank a cup of coffee and put on a peasant's blouse: a true Russian. And what strength that giant had ... With his imperial thumb he bent a silver rouble, and with his shoulders he held up the roof of a restaurant-carriage that had collapsed as the result of a derailment caused by a revolutionary *attentat*, so saving his family, without appearing to be the least put out by the experience.

For the Tsarevich, Alexander III remained 'the example of a father beyond compare'. When he died in 1894, from an attack of nephritis which carried him off in a few months, Nicholas had just turned twenty-six.

Nicholas had been brought up in the English way: sport, modern languages, sport, deportment, *savoir-vivre*, sport, dancing, horse-riding. A well-made body, a head with not too much in it. To be sure, Klyuchevsky, undoubtedly the most distinguished of Russian historians, had taught him the history of former Tsars, but that legendary past was not connected with Russia's current problems. As for literature, he knew little of that, and the taste for it came only later. He had a princely education, but he learned nothing of his trade as monarch.

His adolescent experience began with appointment to the rank of major in the Preobrazhenskys, an élite regiment attached to the Tsar's personal guard. All officers of this illustrious regiment were destined to achieve the highest honours. According to the historian Grunwald, Nicholas II's biographer, the main difference between the Preobrazhenskys and the other great regiment, the prestigious Hussars, was that its drinking sessions were less notorious. The Preobrazhenskys were essentially devoted to horses and women, and were reputed to be the most meticulous in their observation of the rules of the service and the most impeccable on parade.

Nicholas was completely at his ease in this milieu. He took part in all those suppers where champagne flowed and from which he returned with a *bashka*, a terrible head. He often went with his fellow-officers to the opera and the ballet, to which his parents had introduced him when very young, so that he became a real connoisseur. As an adolescent he had among his teachers César Cui, one of the 'Group of Five' composers, who taught at the School of Artillery and communicated to his pupils his passion for harmony. Nicholas II's father had encouraged the rise of native Russian opera, honouring Tchaikovsky, who was given a state funeral, and causing a monument to be raised to Glinka. The enthusiasm for opera was thus developing during Nicholas's teenage years; more than sixty companies were performing in Russia in 1890.

Young Nicholas derived from these spectacles a certain conception of monarchy, of his own sovereignty, of the greatness of the Russian nation, of the patriotism of the peasants. For example, in Glinka's *A Life for the Tsar*, an opera he saw many times, the action takes place in 1613. A peasant, Ivan Susanin, succeeds, at the cost of his own life, in leading into a forest labyrinth a Polish army that is seeking to find and kill the first Romanov Tsar.

The Tsarevich adored Glinka, Tchaikovsky and Mussorgsky, and the diary[2] that he kept until his abdication proves how strong was his passion for the opera. In the month of January 1890 alone he spent, he notes, sixteen evenings at the opera, seeing successively *The Inspector-General*, *Ruslan and Liudmila*, *Mademoiselle Eve*, *Boris Godunov*, *Evgeny Onegin*, *Mephistopheles*, *La Révoltée* and *Lena*, among others.

---

2. In reality it is not so much a diary as a sort of record of appointments, written up afterwards, the insubstantiality of which has contributed not a little to shape the image of an Emperor who was indifferent and weak.

The ballet was also enjoyed by Nicholas no less than by a public who were familiar with the inmost secrets of that art. The quality of the silences and the murmurs of the spectators in the Marynsky Theatre in St Petersburg made it the magical centre where the Russian school established its primacy. This was owed to a large extent to Marius Petipa, who reigned absolutely over the ballet until the coming of Diaghilev and Isadora Duncan in 1908. He was able to pass on to Russia the inheritance of the academic tradition and virtuosity of the Italian school, which aroused wonderment in ballets so different as *The Sleeping Beauty*, *Swan Lake* and *Coppélia*. High society came, mute with pleasure, to watch his creations, and Nicholas attended rehearsals.

When Petipa arrived on the stage, everything was ready, for he never improvised ... He did not look at his dancers as he demonstrated to them the movements and steps, accompanying his gestures with instructions given in frightful Russian: 'You on me, she on you, me on you.' In ten years that was all he had managed to say in the language of Tolstoy. Nevertheless, before assigning roles, he insisted on discovering the private life of his dancers, male and female: 'Have you loved?' he asked one of his ballerinas one day. 'Yes,' she answered, blushing. 'Have you suffered?' 'No,' she answered, laughing. 'Then you shan't dance Esmeralda. To dance Esmeralda well you must have suffered for love.'

One of his most dazzling galas was the one that Nicholas II had him put on for the visit of Wilhelm II, which was held at Peterhof, that mini-Versailles, near one of the lakes that surround the palace. The daises were set up in an arc around the lake, while the stage was placed next to the water. The orchestra was somewhat further off, in a huge gilded cage. A little further off still, an island of rockery had been created, with a grotto from which the dancers came rushing out. The guests arrived in boats all lit up with electric light, which made the décor look like fairyland. As *Thétis et Pelée*, with music by Délibes and Minkus, began, the dancers emerged from the grotto, but an arrangement of mirrors made it seem that they were dancing on the water. The effect was wonderful.

It was at another of these performances that Nicholas met his first love. He went with his father, mother and uncles to the ballet school's graduation ceremony, on 23 March 1890. The most striking of the young dancers was named Mathilde Kshesinska, a girl who came from a theatrical family. She had a great success in the *pas de deux* from *La Fille mal gardée*. Afterwards the Tsar asked in his booming voice: 'Where is Kshesinska?' She was brought to him, curtsied and kissed the Tsaritsa's hand, 'as etiquette demanded'. The Tsar said to her: 'Be the glory and the adornment of our ballet!' 'I felt the Emperor's words as a command,' she tells us in her memoirs. When the guests sat down with the dancers for a banquet, Alexander beckoned her to sit next to him, but the place was already occupied, by a boarder at the school. Kshesinska explained that, as a day-girl, she had no special place to sit at the table. The Tsar then gently moved his neighbour away, so as to make a place beside him for Kshesinska. 'Then he turned to the Tsarevich and told him to sit in the next place, adding on our behalf, with a smile: "Careful, now! Not too much flirting!" '

They fell in love with each other there and then. The idyll lasted several years. But at last came the parting; the heir to the throne had to marry a princess of his own rank. The Tsarevich said to Mathilde: 'Whatever happens to my life, my days spent with you will ever remain the happiest memories of my youth.' He told her that she could go on seeing him and could continue to address him in the familiar way, and that she must always appeal to him if in need. In Russia, scenes of farewell are special moments, precious memories. Of their meeting before the end of their affair, Mathilde relates:

We agreed to meet on the Volkhonsky highway, near the barn and some way off the road. I came from the town by carriage; he rode there from the camp. As always when there is too much to say, tears tighten one's throat and stop one finding the words one would like to utter ... When the Tsarevich departed for camp I remained by the barn and watched him go until he was no longer in sight.

With a view to ending this liaison and also so that Nicholas should learn what the world was like, Alexander III sent his son off on a long voyage, via India to Japan. However, an attempt on his life put a stop to the Tsarevich's travels: a Japanese fanatic struck him over the head with a sword. The Tsar and Tsaritsa, greatly concerned, waited with impatience to hear their son's account of what had happened – but when he got back home he went straight to call on his mistress. When he was in Japan, however, he must have almost forgotten her, since, according to the Japanese police reports, the Tsarevich and his companions spent their nights in the places 'where sailors usually go'.

Nicholas was also fond of simple pleasures. At Darmstadt, for example, where he paid an official visit, he spent his day with the Prince of Hesse, seated at an open window, throwing apples to a crowd. Again, he devoted a whole morning to teaching his dogs to retrieve sticks. Above all, though, he liked hunting. Radzig, his devoted manservant, made several attempts to stop him from giving up entire weeks to this activity. Work had to wait, even the reports of the Minister of the Interior being put aside. One day in 1901 he went off hunting with his minister Sipyagin, and both of them forgot all about a meeting of the Council they had been due to attend. Nicholas is inexhaustible on the subject of these days spent hunting. In his diary for 1892 he writes:

At eight o'clock I set out in a sleigh with Sergei, to fetch Papa and his guests to go hunting. The weather was excellent for our purpose: the thermometer showed zero and there was no wind. We began with the pheasants, behind the coach-house, where there was a terrible fusillade. We went through all the parks until we reached the Lytsie Bugry pavilion, where we lunched. The *battue* ended in Gorwitz Wood at three. Never before have I seen such huge heaps of grouse, 80 or 100 of them. 667 creatures were killed, for 1,596 gunshots. For my share I received 17 birds and 20 hares. When we returned to the house we had tea. Mama had gone to Pavlovsk, returning at 6.30. To dinner we had the Sheremetevs, Mademoiselle de Lescailles, Kutuzov and P. A. Cherevanin.

The imperial hunts took place in the state-owned forest of

Belovezhskaya Pushcha. In accordance with a tradition that went back to the time when this region belonged to the Kings of Poland, the forest was protected from woodcutters: by an *ukaz* of 1803 Alexander I had forbidden anyone to take an axe into it. In 1860 Alexander II had had a small palace built there. In 1889 the fauna were renewed, and, to bring in new blood, stags were introduced from Siberia – but the elks then left, because they could not stand the smell of the stags.

The palace was decorated with the heads of bison killed by the Emperors, while the walls of the dining-room, large enough for 150 people to sit down at table, were hung with stags' antlers, the finest of which, with twenty-eight tines, had been taken from an animal killed during one of these imperial hunts.

As soon as they arrived at the palace the servants brought the Tsar bread and salt, while the Empress, in accordance with tradition, engraved the date of their arrival with a diamond on one of the windows of the balcony. The apartments comprised forty rooms, which were well lit in summer but gloomy and frightening from October onwards. One heard all around the belling of the stags and the howling of thousands of other animals, which alarmed even the toughest. The Emperor arrived in September, accompanied by his General ADC, His Serene Highness Prince Golitsyn and his chief huntsman, in carriages drawn by two horses. (Later, Nicholas was to arrive in a car.) The sportsmen followed a path through the forest until they reached the line from which they were to shoot, where each took up his position, marked by a number. The Emperor's stand was in the centre, with the two best shots on either side of him, and behind them servants bearing the guns. At a determined moment the officer in charge of the hunt gave the order to begin, with a piercing blast on his horn. At once the forest echoed with shouts, crackling sounds and other noises as the beaters advanced, while mounted huntsmen watched to ensure proper procedure. The sounds drew nearer as keepers jerked up and down the rope-barriers they held, with little flags attached. Gradually hemmed in, the creatures rushed towards the sportsmen. That day it was the stags which hurtled forward ahead of the rest, smashing down everything in their path. Then came the boars, in single file, snorting loudly. 'Next, we heard an extraordinary sound, as a single bison traversed the sector at lightning speed, its head down and tail up, followed by two more, then three more ... On this occasion the bison were not shot at, because, owing to an epidemic that had broken out among these creatures, orders had been given to spare them.' The sportsmen had killed stags and boars. Nicholas, who was a good shot, killed five stags. That evening, after taking a glass of madeira, they set out all the antlers, wreathing them with greenery. Then pitch was lit in two urns and a brass band played pieces of music appropriate to each beast: cheerful in honour of the hares, solemn for the bison, graceful for the stags.

In order to ensure the success of days like these spent at Spala, another reservation, a military detachment was sent there, consisting of a lancer regiment of the Guards, the Grodno regiment of hussars, a *sotnia* of Kuban Cossacks and a battalion from the Third Division of the Foot Guards. These soldiers acted as beaters. No one was ever brave enough to protest against such use being made of them.

The Tsarevich would no more have missed a military parade than he would have missed a shooting party. He was fascinated by the parades he attended with his father. A lieutenant remembers one of these:

Music announces the arrival of the Emperor ... We see a brilliant group approaching. At their head rides His Majesty: powerfully built, he seems all of one piece with his superb mount. Here is Russia's strength – here our great, powerful Russia ...

The Tsar is a few paces from us ... With his clear, bright eyes he looks straight into ours and favours us with a smile that is a precious caress. We are dizzy with happiness. We hear: 'Zdravstvuite, Pavlovtsy!' ('Greetings, Pavlovtsy!') We respond with enthusiasm, a resounding 'Hurrah!' bursting from our throats. The band plays: 'God save the Tsar'. We shout, shout in a paroxysm of ecstasy. Our 'Hurrah' is taken up by the other troops ... The Emperor could have asked us to do anything at all, an indescribable feeling had taken possession of us: had he ordered it, we would have jumped into the Neva ...

I believe that on that day there was nobody in the camp at Krasnoe Selo happier than we were.

When he became Tsar, Nicholas was to increase the brilliance and glamour of such parades as this. In Russian tradition these ceremonies served to perpetuate the communion of the Tsar with his people, defence of the native soil constituting, together with anointment by the church, the twofold consecration of his legitimacy.

This identification of the Tsar with his army had, of course, its seamy side, like the identification of patriotism with love for the Tsar. Military defeat might lead to contestation, to revolution. Besides which, the peasants' hatred of military service, which was so lengthy and severe, might also turn against the Tsar. In the reigns of Alexander II and Alexander III anything and everything was done to avoid that service, especially amputation of the index finger of the right hand, the trigger finger. As year followed year the number of wretches who mutilated themselves in this way increased, even though they knew that it meant imprisonment in a fortress, or hard labour. Leo Tolstoy describes these enlisted peasants in *Polikushka*, which was later turned into one of the first great films produced under the regime that began in October 1917. It shows the sad lot of the serf, his departure for the army being seen by the village as a stage on a journey to death. 'Broad is the way that leads to war, narrow the way back,' says the proverb. When Nicholas II's reign began, recruitment to the army was still regarded as a natural catastrophe, a sort of disease which, like an epidemic, signified divine punishment.

For the Emperor, though, this army that marched past him was a token of pride. Who could imagine that it was destined to break in pieces, that the image of these powerful battalions marching on parade would be replaced, within twenty-four hours, by those other images of February 1917, filmed by Pathé-Cinéma, in which the joy of freedom explodes? Those pictures destroy, for good and all, the image of the imperial eagle, the unity of the army with its Tsar. In 1895 what sovereign would have guessed at such a future? The nobility sought to fill the highest posts in the army, and, in 1914, 87 per cent of the generals were still of noble origin, despite the efforts made by the Military Academy to bring forward the ablest officers, regardless of social background. During Nicholas II's reign the proportion of nobles among the officers fell from 72 to 51 per cent, but this movement had not reached the top ranks of the hierarchy by 1914, because, under pressure from those close to him, the Tsar slowed it down. One consequence was that a section of the officer corps embraced liberal ideas, a development that was itself not without consequences.

Amusing himself, attending military parades (sixteen in a single month), dining in fine company, Nicholas loved all that. But rumour swelled and stories circulated. It was said that his father had put him under house arrest at Livadia because he thought of nothing but his ballerinas. It was said, too, that he was unwilling either to marry or to reign. It was said, even, that his father was thinking of making Nicholas's young brother Grand Duke Michael his heir, instead. It was said ... But what is not 'said'? About Louis XVI and Marie-Antoinette, too, many things were said, and scabrous stories were made spicier by depicting this royal couple, along with the Comte d'Artois, as gay and shameful participants in licentious adventures. It was a sign that the monarchy no longer inspired respect. In the same way, when the Tsarevich came home from his visit to Japan, people said that the man who had attacked him was a husband infuriated by Nicholas's persistent advances to his young wife.

On 25 February 1892 Nicholas recorded in his diary:

Two days ago I was made a member of the Finance Committee: *much honour but little pleasure.* Before the meeting of the Council of Ministers I received six members of this institution, the existence of which, I must admit, I had never suspected. We were in session for a long time, until 3.45, and that made it too late for me to go to the Exhibition.

On another occasion he wrote: 'An insufferable morning. I received the Swedish minister and the Japanese monkey.' The 'Japanese monkey' was the envoy of the Mikado; Nicholas called all Japanese 'monkeys'. He mistrusted the Poles, despised the Jews, but was confident of the loyalty of his Muslim subjects.

S. Witte, Alexander III's minister who tried to modernize the country's economy, had put it to the Tsar that the Tsarevich's training would be helped if he were made chairman of the committee for building the Trans-Siberian Railway.

'What?' replied the monarch. 'But you know the Tsarevich well. Have you ever had a single serious conversation with him?'

'No, sire, I have not had the pleasure of such a conversation with the heir to the throne.'

'Why, he is a mere child. He has only childish notions. How could he preside over that committee?'

Gustave Lanson, author of *Histoire de la littérature française*, taught French to Nicholas for a time, and had no reason to complain of his pupil, quite the contrary. The Tsarevich even had a taste for the French classics – Molière especially, La Fontaine, Mérimée. All the same, the qualities and the docility that are needed for mastering foreign languages – and Nicholas showed himself gifted for learning French, German and English – do not necessarily help when it comes to understanding the world around us. And on that point Pobedonostsev held the same view as Alexander III. When Pobedonostsev tried to explain to the Tsarevich how the state functioned, 'he became actively absorbed in picking his nose'. All accounts agree: as soon as problems of government were raised, Nicholas, like Louis XVI, was suddenly overwhelmed with an inexpressible torpor. Nevertheless, through persistence, Pobedonostsev did succeed in getting some of his ideas into Nicholas's head, and especially this one: that it was the duty of a Tsar-autocrat to pass on all his powers *intact* to his son.

The contrast is indeed striking between the Tsarevich, graceful and polished, always shining and well groomed, and the Procurator-General, dressed all in black, with his round spectacles and his bow-tie, who looked as though he had stepped straight out of a novel by Dostoevsky. In him the flame of the autocracy never went out, and he argued with a talent and a verve that gave brilliance to his austere appearance and his burning eyes.

Pobedonostsev's arguments were powerful and coherent enough for even a pupil as inattentive as the Tsarevich to retain their outlines in his memory. He met head-on those who spoke of Russia's backwardness and said that she was not 'ready' for democracy. In his view, the introduction of democratic practices would actually mean a step backwards. In other words, Pobedonostsev did not criticize the tempo of reform but its very principle.

Above all, he criticized the parliamentary system, 'the great falsehood of our time'. 'The fine ideas of the Great Century were propounded by élites, but, with the coming of universal suffrage, the way they were applied became dependent on an uneducated, uncouth majority.' The example of the French parliament was there, at the time of the Panama scandal, to demonstrate that 'the public good, to which the people's deputies were supposed to be devoted, was merely a sort of alibi for politicians who were corrupt and irresponsible'. Moreover, parliamentarism resulted in the election not of the best men but of the most demagogic. His third argument was particularly relevant to the case of Russia: in a country of many different nationalities, elected representatives would defend the interests of their own communities, 'each for itself', whereas a monarch could alone embody the *common* interest.

In a period when the political press was developing very rapidly, Pobedonostsev stigmatized journalists, whose claim to speak for public opinion he contested. Under cover of the alleged common interest, these men sought primarily to advance themselves, even if in doing so they destabilized a society about which they cared nothing. They acted irresponsibly, rather like members of parliament, or like those schoolmasters who not only fail to carry out their function of teaching the three Rs but also fail to instil 'the duty of knowing, loving and fearing God, of loving our native land and of honouring our parents'.

Unappreciated in the West, Pobedonostsev was one of the fathers of a reactionary, traditionalist body of ideas which he was able to group around two or three themes. All these ideas were familiar to Nicholas; he had absorbed them and they became part of his view of the world. The Tsar's authority was sacred because it came to him from God; the Russian people were fundamentally good; the intelligentsia were breathing evil into the people, evil that was rising up on every side.

Two or three days before the attack of nephritis that carried him off, Tsar Alexander III had had a violent quarrel with his son. Nicholas had fallen in love with the sister of Wilhelm II, whom the Tsar loathed. He wanted his son to marry Princess Alix of Hesse. But the heir to the throne hung back. Alix was a head taller than he was, and he was very sensitive about his short stature – short, that is, in relation to some giants among those around him, for he was himself 1.70 metres (about 5 feet 6 inches) tall. Then, however, his attitude underwent a sudden change, and he became infatuated with the haughty princess. There was still another difficulty to be overcome, though. Alix was a Protestant, and devoted to her religion, whereas if she were to marry Nicholas usage required that she convert to Orthodoxy. Wilhelm II urged her to abandon the faith in which she had been brought up, his intervention being motivated by hope to win favour thereby with the Tsarevich, and so to restore the 'Three Emperors' Alliance' that Alexander had broken and, consequently, to nullify the agreements that the Tsar had signed with Félix Faure in 1891. Eventually, though not without reticence, Alix agreed.

This bringing together of frivolity with prudery transformed the fickle Nicholas into a faithful lover. On 8 April 1894, the day of his betrothal, he wrote in his diary:

A marvellous, unforgettable day. Today is the day of my engagement to my darling, adorable Alix. About ten she came to Aunt Miechen's and after a talk with her we came to an understanding. O God, what a mountain has rolled from my shoulders ... The whole day I have been walking in a dream, without fully realizing what was happening to me. Wilhelm sat in the next room and waited with the uncles and aunts until our talk was over. Then we went to the Queen [Victoria] and after that to Aunt Marie, and there was kissing all round. After lunch we went to Aunt Marie's private chapel for a thanksgiving service. Next morning the Queen's Dragoon Guards performed a whole programme under my windows, a touching gesture. At ten my wonderful Alix arrived and we went together to have coffee with the Queen ... Then we went off together in a little carriage to Rosenau. I drove, what joy.

When Alexander III succumbed to his mortal illness, Alix was summoned to his bedside, in her capacity as consort to the Tsarevich. The Emperor had desired to wear his finest uniform, for the last time, to fit the occasion, but the Princesses Obolenskaya and Countess Vorontsova received the German girl with coolness. Alix noted this unfriendly attitude, and also, and especially, that of the soon-to-be-Dowager Empress Maria. She confessed later that, sad and alone, she prayed to the Lord to reconcile her with her mother-in-law.

When the Tsar died there was the seemingly endless funeral journey, following the coffin, from Livadia in the Crimea to St Petersburg, with services, again and again, in every town on the way, to do homage to the late ruler. 'A bird of ill omen,' Russians said, crossing themselves, when for the first time they saw the fiancée of the new Emperor, walking all in black behind the bier. 'That was how I entered Russia,' Alix observed; 'and my wedding, held immediately afterwards, seemed to me another of those funeral services I had just been attending.'

In Russia the Tsar's wedding, though accompanied by festivities, was not associated with such cheerful memories as in other countries, France for instance. Over there, instead of jokes about the love-life of Henri IV or Louis XIV, what the people remembered was that their monarch's bed was often the source from which sprang wars, either civil or foreign, or that his queen suffered a sad fate. Russia's woes thus remained linked with the wedding ceremony of her ruler, so often tragic. Were the new couple conscious of that? Alix had already sensed that she was marrying the Tsar in a distressful context. As for Nicholas II, he was mainly concerned with the changes in the relevant protocol that had been made since Gregor Kotoshikhin[3] gave his description of it in 1666, an account well known to the Tsar, even though some parts of it had fallen into desuetude.

When the Tsar is crowned he is at the same time anointed with holy oil, and so becomes the Lord's Anointed. After this the high authorities, the boyars, his closest advisers and persons of all positions in the state congratulate him. And when they congratulate him the Patriarch, the Metropolitan, the Archbishops, the Bishops and other high clergy bless the Tsar with icons and offer him bread, velvet, moiré, satin, silk both gilded and plain, sables and silver vessels.

When the Tsar is to marry, thought has to be given to the roles in the ceremony that must be assigned to the great men of the realm and their wives, who are to be in the wedding procession, who will be master of ceremonies, the placing of the best man, the matchmakers, the candle-holders, the bread-bearers, and so on. The programme having been drawn up, it is sealed and signed, and an *ukaz* by the Tsar announces the day when the wedding is to take place; he then sends out orders to all concerned to be ready to play their part in the ceremony. 'And if, during the wedding, anyone shall cause trouble through disputes over rank, he shall be put to death, without mercy, and his property confiscated. Similarly, if, after the wedding, anyone shall abuse another on account of the part assigned to him in the ceremony, it shall bring upon him grave disfavour and punishment by the Tsar.'

'The functions to be performed by the participants in the wedding ceremony are these. On the Tsar's side, first and foremost, as sponsors, a father and mother, as proxies for the Tsar's own parents.' Next, in procession: an archpriest carrying a cross, the master of ceremonies, the Tsar himself and eight boyars. These persons 'go with the Tsar and Tsaritsa to the wedding in the church, and at table sit above all the rest'.

3. An under-secretary in the diplomatic department in the time of Alexei, the second Romanov Tsar (1645–76).

The task of the best man and his pages is to summon the guests to the wedding, and at the wedding to deliver addresses on behalf of the master of ceremonies and the Tsar and to distribute gifts.

The task of the bridesmaids is to dress the Tsaritsa and look after her, putting on and taking off her ornaments. While the Tsaritsa is being dressed the candle-bearer stands by, holding a candle. The task of the bread-bearers is to bring bread to and from the church, carrying it on litters that are lined with gilded velvet, and covered with embroidered cloths and sables ...

Then there are twelve boyars and their wives who sit at table with the Tsar's mother and father but do not go into the church with the Tsar.

Finally, the butler is stationed by the sideboard, in charge of the food and drink ...

On his wedding morning the Tsar goes to the cathedral to pray. While he is praying the Patriarch blesses him with the cross and sprinkles holy water on him ... Then the Tsar goes to another church, the one where his predecessors are buried ... [There he prays before returning home.]

The palace is hung with velvet tapestry, and large carpets, from Turkey and Persia, are laid on the floors. They arrange the place where the Tsar and his bride are to sit, and set before them a table, and also tables where the boyars are to sit. These are covered with cloths, and bread and salt are placed thereon.

The Tsar puts on his coronation robes. The bride is dressed like a Tsaritsa, except for the crown, wearing on her head instead a maiden's garland ... The Tsar orders his bride's parents and all their side of the wedding party to go with her to the palace and there await his arrival ... When the Tsar has been informed that they are all in the palace, he tells the archpriest who is his confessor that the time has come to go: the confessor says a prayer while the Tsar and his companions pray before the icons ...

The Tsar and his retinue enter the palace to find the future Tsaritsa standing there with her people.

The Tsar and his bride sit down on the same cushion, and then all the rest sit down in their places at table, and food is brought to them. Before they eat, the Tsar's confessor says the Lord's Prayer.

The pages and bridesmaids ask the blessing of the bride's father and mother before dressing the bride's hair. The confessor and the wedding guests begin to eat and drink, not in order to satisfy hunger but because the ritual requires this. Meat is set before the Tsar and carved, but he does not touch it until they have finished dressing the bride's hair.

While the hairdressing proceeds, the Tsar and his bride are hidden behind a veil, which is held by the candle-bearers.

The hairdressing finished, a cross is sewn on the veil and the bride's garland removed. Gifts are distributed. After the third course, the confessor rises and says grace. Then they go to the church, where the bride's parents 'bless the Tsar with icons decorated with gold, precious stones and pearls. After which they take their daughter by the hand, place it in the Tsar's hand, and withdraw.' As they leave the church, 'all the bells start to ring and in all the churches prayers are said for the health of the Tsar and the Tsaritsa and for their union in lawful wedlock.'

Unlike what happened in those distant days, there was no reception after the wedding, because of the court mourning for Alexander III, and the ceremony took place not in the Kremlin but in St Petersburg.

This is the report of the Tsar's wedding sent to Paris by the correspondent of *Le Matin* in a dispatch dated from St Petersburg, 26 November 1894:

At dawn a salvo of twenty-one guns fired from the fortress announced the celebration of the Emperor's wedding. The actual ceremony began at 12.45, when the cortège set forth.

The vast halls of the Winter Palace present a dazzling scene, with the variety and brilliance of the uniforms, gleaming with gold and silver, worn by the crowd of officers, high dignitaries and marshals of the nobility.

The Empress' many ladies and maids of honour wear court dresses of velvet, garnet-coloured, with, on their heads, that sort of Russian crown called a *kokoshnik*, also made of velvet, and embroidered with pearls. A long white veil falls from the *kokoshnik*, forming a long train which the ladies carry. The dress, which is fronted with a white apron, has long, flowing sleeves attached at the shoulders. It is low-cut and embroidered with gold.

The society ladies who are allowed to attend the court also wear the court dress, but many of these are in silk of various colours, studded with diamonds.

The Empress Dowager wears a more austere form of court dress, made of white wool, with a long train borne by four chamberlains.

The bride wears a wonderful costume which emphasizes her beauty and youth – a Russian court dress in white silk embroidered with silver, on her head a diamond diadem, her shoulders covered with the Imperial mantle of gold brocade lined with ermine, and a long train borne by four dignitaries, two on each side, and the Grand Chamberlain, who holds the end of it.

The Emperor wears the red uniform of a colonel of Hussars of the Guard, with a pelisse hanging from his shoulders.

The Prince of Wales and the Duke of Coburg are in Russian uniforms, the Duke of York in naval uniform. All the princes wear the ribbon of the order of St Andrew.

Entering the chapel, the King of Denmark, who gives his arm to the Empress Dowager, is followed by the Emperor, who accompanies his bride. Then come the Grand-Duke of Hesse with the Queen of Greece, the Duke of Coburg with the Princess of Wales, the Prince of Wales with Princess Irene of Prussia, the Prince of Romania with the Grand-Duchess Vladimir, Prince Vladimir with the Grand-Duchess Sergei, the Duke of York with Grand-Duchess Elizabeth and so on.

During the marriage ceremony the Grand-Dukes hold crowns over the heads of the bride and bridegroom, in accordance with Russian custom. Pairs of Grand-Dukes succeed each other in this role, each one holding a crown very high above the head of each of the couple, so that all the Grand-Dukes take part in the ceremony, which is very impressive.

After the nuptial blessing and before the Te Deum a touching scene is observed: the Emperor and the new Empress approach the Empress Dowager, to salute her. The Empress Dowager kisses them warmly, and blesses them in her own name and in that of him who lies in the church of the fortress. This is a very emotional moment. We notice, too, how affectionately the King of Denmark embraces the newlyweds.

The wedding concludes shortly before two o'clock. When everything is over at the palace, the Empress Dowager, in accordance with protocol, steps into a court carriage drawn by four horses, with two Cossacks on the back seat, and sets off for the Anichkov Palace, along the Velikaya Morskaya and the Nevsky Prospekt.

Five minutes later, the newlyweds go to the Cathedral of Kazan in a four-horse carriage with postilions. Both of these carriages are closed.

Along the route a close-packed crowd presses from both sides. The windows are open and full of people, as are the balconies. Ladies wave handkerchiefs. A formidable rumbling of cheers accompanies the carriages of the Empress Dowager and the newlyweds.

The new Empress has a grave and impressive manner. As she passes in her carriage, along with her husband, after the ceremony, I see that she is smiling, her cheeks pink, happy to hear the crowd's ovation.

The Empress Dowager looks very serious, with a touch of sadness in her face which is sometimes banished as she thinks of her son's happiness. She seems, in any case, sufficiently recovered from the dreadful trials and sufferings of the last month.

When the newlyweds arrive at the Kazan Cathedral they are received by Palladius, Metropolitan of St Petersburg, who, surrounded by the clergy, awaits them with cross and holy water.

The crowd is immense, so that the Imperial carriage has difficulty in getting to the steps of the cathedral. The cheering redoubles, drowning the sound of the bells and that of the fortress cannon.

The Emperor and Empress enter the cathedral, where a solemn Te Deum is celebrated, after which they devoutly kiss the miracle-working icon of Our Lady of Kazan.

As Their Majesties emerge, a ray of sunlight suddenly touches them. The thunderous cheering is then renewed and accompanies them to the Anichkov Palace.

For a prince so pious as Nicholas the coronation, his marriage to Russia, was vitally important. His piety was not merely innate, it was one of his attributes. If he could not become a saint, like Russia's first princes, he must be pious, because, for the Tsar as for every Russian, evil was to be judged in relation to piety. The anointment at coronation, an old tradition inherited from Byzantium, washed away his sins.

Moreover, this ceremony took place in Moscow, a circumstance that helped to perpetuate the pre-eminence of the Third Rome over St Petersburg. The latter city, gateway to the West and the country's new capital, was not the heart of Russia. Peter the Great had indeed made it the embodiment of his empire's power, but Holy Russia did not feel at home amid those Venetian palaces. Besides, the imperial city had broken the link of filial love that bound it to its monarch; Alexander II had been murdered there.

Alexander III had rejected St Petersburg and preferred Moscow. Nicholas II felt ill at ease in the new capital, and found pleasure only in visiting the army's training ground at Krasnoe Selo, nearby.

It was in Moscow that, through the coronation ceremony, the union between the sovereign, his state and the church found expression, and in Moscow that he incarnated the pious Tsar of tradition. There, too, were buried his predecessors on the throne. On a number of occasions during his reign Nicholas expressed the desire to spend Holy Week in Moscow. There it was that he commemorated the victory over the Tartars, and also Borodino, the victory over Napoleon. During the coronation proceedings, just as later, in 1913, when the tercentenary of the dynasty was celebrated, Moscow's fervent greeting to its rulers confirmed his feeling for the city. This preference had, furthermore, a political significance. Moscow incarnated tradition, Rus, the Russia of olden times, as against Rossiia, the empire. Pobedonostsev was not mistaken when he called his collection of political reflections *Moskovsky Sbornik*, a 'Muscovite' collection of writings.

While the coronation festivities were being prepared in Moscow, it was no accident that Nicholas II and Alexandra withdrew to pray in the Stone Palace, outside the city, as though to purify themselves. When their son, the heir to the throne, came to be born, he would be called Alexei, after the most pious of the Romanov Tsars.

Nicholas II preferred Moscow to St Petersburg because the old city embodied the past, whereas St Petersburg represented modernity, the Enlightenment and atheism. All this was not explicitly formulated in his mind, yet there was obvious coherence between his opinions, his conduct and this political action. Since that coherence was not given expression in the form of ideas or concepts, people assumed that he was empty-headed or weak and subject to influence – first from his mother, then from his wife. It was said, too, that the last councillor to speak with him was the one who had the last word. Actually, the one who had the last word was the one whose opinions corresponded to the Tsar's. Otherwise, Nicholas bent without saying anything and without blinking.

Whenever he could, Nicholas dressed in the Russian style, in a blouse, and spoke Russian, using English and German only within his family. He was fond of his native tongue, and liked to quote from Pushkin. Among Russian writers he particularly enjoyed Gogol – for whom

in St Petersburg everything breathes falsehood ... That Nevsky Avenue lies all the time, and especially when the night spreads itself over it in a compact mass ... It is a town without character: the foreigners who have grown fat there no longer look like foreigners, while the Russian inhabitants have become somehow foreign, and are no longer either one thing or the other.

Nicholas II did not like the founder of St Petersburg, even though Klyuchevsky, who taught him history, was not hostile to Peter the Great. 'I recognize my ancestor's great merits,' the Tsar said to his doctor one day, 'but he is the ancestor who appeals to me least of all. He had too much admiration for European culture ... He stamped out Russian habits, the good customs, the usages which are the nation's heritage.'

To Peter the Great he preferred Peter's father, Alexei the Pious (1645–76), the most traditionalist of all the monarchs who ever ruled over Russia. His 'model' was a gentle man who was too weak with those around him. His advisers robbed him but the people rose up and got rid of them for him. Alexei was victorious over rebels, in particular Stenka Razin, a Cossack leader who plundered the coasts of the Caspian Sea and created a revolutionary mood in that region, stirring up the peasants against the landlords and officials. He was caught and executed in 1671. Another success of Alexei's was the 'historic' choice made in his reign by the Ukrainians, at their assembly at Pereyaslavl, when, having to decide between accepting the suzerainty of Turkey, Poland or Russia, they opted for the last.

This prince was, like Nicholas, very pious, and it was in his honour that the Tsar organized the grandest costume ball of his reign, in 1903. The entire court dressed up to recreate the atmosphere and ways of Alexei's time. This love of the past, this conservatism, is observable in all the Tsar's tastes – for instance, his love of old icons, or his attachment to traditional forms of spelling. The question arose of improving Russian spelling by doing away with the 'hard sign', which had to be written at the end of a lot of words, though it performed no particular function. A dispute over this matter divided writers, grammarians and teachers and became an affair of state. 'Personally,' Nicholas told Dr Botkin, 'I shall never trust nor give a responsible position to a man who omits the hard sign.'

Before the coronation there was unusual activity in the city for a period of several weeks, because the ceremony would last for five days and over seven hundred guests were expected to attend the various receptions to be organized on this occasion. Persons who hoped to be present at the festivities arrived on foot from all over Muscovy. Thousands upon thousands of gifts would be distributed.

At his coronation, Nicholas would be proclaimed Emperor and Autocrat of All Russia. His style thereafter would be:

'We, by the Grace of God, Nicholas Alexandrovich, Emperor and Autocrat of All Russia; Tsar of Moscow, Kiev, Vladimir, Novgorod, Kazan, Astrakhan, Poland, Siberia, the Tauric Chernonese and Georgia; Lord of Pskov; Grand Prince of Smolensk, Lithuania, Volhynia, Podolia and Finland; Prince of Estonia, Livonia, Courland and Semigallia, Samogitia, Belostok, Karelia, Tver, Yugra, Perm, Vyatka, Bulgaria and other lands; Lord and Grand Prince of Nizhny-Novgorod and Chernigov; Ruler of Ryazan, Polotsk, Rostov, Yaroslavl, Belo-Ozero, Udoria, Obdoria, Kondia, Vitebsk, Mstislavl and all the Northern Lands; Lord of the Iberian, Kartalinian and Kabardinian lands and of the Armenian provinces; Hereditary

Lord and Suzerain of the Circassian and Highland Princes and others; Lord of Turkestan; Heir to the Throne of Norway; Duke of Schleswig-Holstein, Stormarn, the Dithmarschen and Oldenburg.'

In a major 'first' for history, the camera fixed for all time the most striking moments of the ceremony. Its pictures show, first, Nicholas's solemn entry into Moscow on 6 May 1896. His white steed is followed by the cortège of notables on horseback, and then by the long procession of guests: Henry of Prussia, Wilhelm II's brother; the Duke of Connaught, son of Queen Victoria; Nicholas of Montenegro; the Crown Princes of Greece and Romania; three Grand Dukes; a Queen; two reigning monarchs; twelve royal heirs; and sixteen other princes and princesses.

At the Cathedral of the Assumption, before the high clergy in all their finery, Metropolitan Sergei read the traditional sacred text. As the Tsar mounted the steps of the altar, the heavy chain of the Order of St Andrew slipped from his shoulders and fell to the ground. Nobody noticed this except the persons following him, and they refrained from drawing attention to it, as it would have looked like a bad omen. After he had received the crown, Nicholas II placed it reverently on the head of the Empress. All the bells of Moscow's 101 churches started to ring in unison. The festivities then began. They ended in catastrophe.

After the processions had passed, the public who had come from all parts of Russia assembled on the Khodynka Fields and pressed towards the platforms for the traditional distribution of gifts. From six in the morning everyone was trying to get into the best positions for this event. Suddenly the crowd rushed forward as though pursued by fire. The rear ranks trampled on those in front in an indescribable scrimmage, and thousands of people were crushed and suffocated. When the people came to their senses a few moments later, the damage had been done. A total of 1,282 corpses were collected from the scene, and the injured numbered between 9,000 and 20,000, according to different estimates.

The government took charge of the funerals of the victims and paid 1,000 roubles to each bereaved family, to the great surprise of

Li Hung-chang, the Chinese Ambassador, who had been invited to the coronation and witnessed the catastrophe. He explained that, in his country, the Emperor would not have taken any note of a disaster such as this.

The Tsar thanked the people for the enthusiasm they had shown for him during the coronation festivities, 'a moving consolation after these days of grief'.

These events were seen as a sinister portent. A few days later a hailstorm of unprecedented violence struck Nizhny-Novgorod while Nicholas II was inaugurating there the first great exhibition of Russian industry. Misfortune was dogging the country.

And yet, as Bernard Pares, visiting Russia for the first time in 1898, was able to observe, Russia's factories could outdo those of Britain or Germany. They produced sophisticated weapons and explosives, using the cotton of Turkestan. The effects of this work had been seen in the campaign in Bulgaria in 1878. At Nizhny-Novgorod, Russia was displaying for the first time the progress she had accomplished.

Another observer made the same diagnosis, namely, Vladimir Ilyich Ulyanov, later known as Lenin. It was in 1896 that he began one of his most important works, *The Development of Capitalism in Russia*.

At the death of Alexander III, as at the beginning of every reign, requests and petitions to the new ruler naturally reached him in great numbers, especially those coming from the zemstvos. These local administrative bodies created by Alexander II constituted an adumbration of a representative regime, even though they were largely dominated by nobles and bureaucrats. The zemstvos brought together doctors, agronomists, vets, teachers, and so on – all the educated men of Russia's provinces. They formed the cadres and nucleus of a decentralized administration, but had seen their sphere of competence reduced since the accession of Alexander III, and even earlier, as is explained by Khizhnyakov, who presided over one of the zemstvos:

Their interminable martyrdom had been going on for thirty years ... Their petitions were met with negative responses, or, rather, in most cases, were left unanswered. More than once, the explanation given for a negative response was that the needs referred to in the petitions were not justified, since other zemstvos had not presented similar petitions. Yet, at the same time, all attempts by the provincial assemblies to concert their efforts were severely repressed.

The Russians awaited impatiently the new Tsar's reply to their petitions calling for an increase in their resources and in their powers. But they were totally unaware of the new Tsar's opinions on the matter. His answer, when given, was very clear:

It has come to my knowledge that, of late, there have been heard in some zemstvo assemblies the voices of those who have been carried away by senseless dreams that the zemstvos might be called to participate in the government of the country. Let it be known to all that I shall devote all my strength, for the good of the whole nation, to maintaining the principle of autocracy just as firmly and unflinchingly as it was preserved by my unforgettable father.

It has been claimed that in this passage, apparently written by Pobedonostsev, the original expression was not 'senseless dreams' but 'baseless dreams'. Even if a slip was made here, the expression corresponded well to the Tsar's actual ideas.

This address of his was a challenge to liberal opinion. It immediately strengthened the radical elements whom Alexander III's repressive policy had neutralized to some degree. Above all, it destroyed the illusions of those persons who imagined that life would change with the coming of the new Tsar, and that fresh possibilities would open up for society. This disillusionment is expressed in Chekhov's play *The Seagull*, written in 1896. In *Three Sisters*, *Uncle Vanya* and *The Cherry Orchard*, written in those same years of disillusion, we see the cultivated society of Russia's provinces, without prospects, and at once unconscious and sensitive, becoming anxious about the rumblings it hears all around it – the peasant world stirring, the capital more indifferent than ever, and itself

incapable of action. The characters in these plays reveal their helplessness and their lack of decision regarding their own future.

Like Masha, the disappointed lover in *The Seagull* who always
wears black because she is in mourning for her life, sad and in
despair, the Russians were in mourning for their history. While they
dreamt of a new life, it seemed to them that time had stopped. They
then asked aloud how their history could be set going again – a
history that they wanted to build but in which they had no role to
play.

And, indeed, nothing changed, because the state made immobility
its principle. Now here was Nicholas II intending to continue this
closure imposed by Alexander III upon the era of reforms ... The
autocracy was held responsible, and the young generations absorbed
the ideas of the nihilists. Herzen, Pisarev and the others questioned
everything: the political system, of course, but also morality, art,
religion, marriage. 'You can't mess about with history and remain
unpunished for it,' said those who watched the changes that had
taken place in France, Prussia, Italy and Spain since 1815. For
them the West was a model, and for P. Struve or I. Petrunkevich
the autocracy was an obstacle in the way of the nation's development. Its historical function was exhausted. The Marxists did not
disagree.

On this point, at least, the 'Westernizers' agreed with the 'Slavophils', who considered that the Russian spirit was incompatible
with reforms conceived abroad. This was an old dispute, dating
from Peter the Great (1682–1725) and even earlier, since Adam
Olearius in the 1630s and Augustin von Mayerberg in 1661, during
their visits to Russia, both made mention of it.

Under Nicholas II one of the Slavophils, Sergei Bulgakov, who
had begun as a Marxist and who belonged to the new generation of
revolutionaries, like Lenin, Gorky or Martov, who were contemporary with Nicholas, shows the gulf that separated the Tsar,
with his pleasures, his ceremonies and his priests, from this intelligentsia in love with the sciences, with medicine, keen to solve the
social question, the woman question and others yet. The description
he gives of the intelligentsia enables us to appreciate the mutual lack

of understanding between two worlds separated by an invisible barricade.

It is well known that there is no intelligentsia more atheistic than the Russian. Atheism is the common faith into which are baptized all who enter the bosom of the humanistic intelligentsia church ... And just as any social milieu elaborates its own customs, its own peculiarities of faith, so the traditional atheism of the Russian intelligentsia has come to be taken for granted, and it is even a sign of good form not to talk about it. In the eyes of our intelligentsia a certain level of education and enlightenment is synonymous with indifference to religion and repudiation of it ... [This atheism] is most often taken on trust, and preserves the characteristics of a naïve religious belief, only inverted. The fact that it assumes militant, dogmatic, scientific-like forms does not alter this ...

Our intelligentsia, which almost to a man strives for collectivism, for a possible communality of human existence, is in its own temperament anti-communal, anti-collective, for it bears within itself the divisive principle of heroic self-affirmation. To a certain extent its hero is a superman, who adopts with regard to his neighbours the proud and defiant pose of a saviour. For all its striving towards democracy, the intelligentsia is merely a special kind of aristocratic class arrogantly distinguishing itself from the philistine crowd. Whoever has lived in intelligentsia circles knows well their haughtiness and conceit ... their scorn for those of other opinions ... It regards socialism itself ... as a supra-historical 'final objective' ... requiring an historical leap by an act of intelligentsia heroism ...

The Russian intelligentsia's cosmopolitanism is also well known. Raised on the abstract schemes of the Enlightenment, the *intelligent* ... regards himself as a citizen of the world. This ... absence of a healthy national feeling is tied up with the intelligentsia's isolation from the people.

The intelligentsia has not yet thought through the national problem ... The modern-day Marxists ... have dissolved it without trace in the class struggle.

The zemstvos which had sent petitions to the Tsar contained *intelligenty* of all sorts – Tolstoyans as well as Marxists, nihilists and Slavophils, heroes both Dostoevskian and Chekhovian. They were mostly members of the upper or lower nobility and landlords. Nicholas II's reaction to the zemstvos was revealing. They constituted the first experiment in 'local self-government'. The provincial élites who composed them included both cultivated families and also uneducated countryfolk who had made money, real peasants who had become speculators and entrepreneurs. They were concerned only with problems of local or provincial administration. Sooner or later, though, they would tackle more general problems, to do with taxation, the infrastructure, and so on, and would engage in politics.

Now, the Tsar considered that such interventions on their part ran counter to the mission which God had entrusted to him, namely, to safeguard the rights of the autocracy. That the question of the state should clash with the powers he possessed seemed to him absurd. The idea that the zemstvos could become representative assemblies made him bristle, because he felt that, stage by stage, such assemblies would come into conflict with the state and the bureaucracy. Pobedonostsev's ideas thus coincided with his own. And the discussions between Westernizers and Slavophils, who differed about the country's future and wondered whether Russia was not part of Europe, or, though part of Europe, was free from Europe's bad features – capitalism, individualism, egoism – irritated and bored him. Actually such discussions were of no interest to Nicholas, but he sensed that all this ultimately threatened the very principle of the autocracy. He loathed these masters of words who isolated him from the people with whom he claimed to be at one.

Nicholas II asked that the word 'intelligentsia' be removed from the dictionary.

After thirteen years of reaction, the Tsar's reply to the zemstvo petition produced a veritable shock wave. So, then, the reign of Nicholas II was to be a continuation of that of Alexander III. All the hopes cherished by Russian 'society' fell in ruins. What was meant in those days by 'society' (*obshchestvo*) was the community of cultivated Russians who embodied the essence of the nation, but from whom autocratic arbitrariness withheld the possibility and right to work for the nation's development and greatness.

During Alexander III's reign this return to the past had stimulated, in response to the 'reaction', the development of a liberal movement within the zemstvos. Its members wanted the Tsar to grant a constitution to Russia, but they 'did not want war with the autocracy ... They preferred a peaceful path, with stages ... an evolution.' Vasily Maklakov, who was one of them, considered that, given the disillusionment caused by the way Nicholas's reign began, the principles of the liberal movement would change and the old strategy would give way to a fight against the autocracy. Thenceforth, leadership passed from the zemstvo men into the hands of the politicians, that is, of a radical intelligentsia.

In fact there were radicals in the zemstvos just as there were moderates in the new intelligentsia, but the zemstvos were no longer able to perform *on their own* the function that society expected from them, and in order to wage the struggle in radical fashion it was necessary to create organizations of a new type, political parties. These parties all appeared just after the Tsar's discouraging address to the zemstvos, between 1896 and 1900. Their common aim was 'national liberation', meaning liberation from the autocracy, after which each would go its own way. The idea of a national union prevailed, a union embracing all tendencies and all classes of society, from the most moderate liberals in the zemstvos to the most revolutionary *narodniki* (populists).

As early as 1894 the Narodnoe Pravo (People's Right) group and the Union of Liberation proposed 'to unite all the opposition groups and organize an active movement that, helped by the moral and spiritual forces which inspired it, would triumph over the autocracy and ensure for everyone his rights as a citizen'. After the Tsar had rejected the zemstvo petition, many zemstvos became radicalized. Peter Struve, one of the leaders of the movement, thought that they would nevertheless continue to constitute the most suitable form of organization, because they covered a large part of the country. He also urged the liberals to meet the revolutionaries, as he himself met Lenin, since, with the growth of industry and the development of the working class, the zemstvos, while certainly remaining one of the main fields of political action, would no longer be the only ones. Even though they were emanations of society, the zemstvos were recruited on a property-qualification basis, and the imperatives of

democracy required a broader social foundation, with a direct appeal
to the popular forces. The urgent need to create these political parties
of the Western type appeared at the moment when, as in the West,
students and workers began to exist and to make themselves felt.

The foundations of society had indeed been changed. The towns
were becoming transformed and modernized. Primary education was
developing and spreading to the countryside, thanks to the zemstvos.
The policy pursued by Witte increased the poverty of a debt-
burdened small peasantry that flocked into the industrial cities and
suffered terrible exploitation there until wildcat strikes broke out.
For Lenin, for instance, the crisis which best symbolized this process
of change was the strike in the Morozov factories in Vladimir
Province in 1885. The cotton industry was in trouble, and the
employer had hoped to save his bacon by increasing the fines payable
by the workers, an indirect method of reducing their wages. A large-
scale strike broke out which enjoyed the sympathy of public opinion,
to the extent that a jury acquitted strikers who had been arrested
and charged.

Thus, for the second time, following the precedent set by the
acquittal of Vera Zasulich, a jury showed its independence: in the first
case, independence of the authorities, in the second, independence of
the rich. The lesson was obvious. Against the system and the 'justice'
that incarnated it there existed that union for which Peter Struve
tried to create a structure. Though he did not succeed, he set many
things going.

The democratic idea was progressing of its own accord, and there
was a coming-together of the liberal elements in the zemstvos –
teachers, doctors, agronomists (the rural intelligentsia) – and the
active elements in the towns who were alarmed at the conditions
endured by the workers. The processes were the same; after the
famine of 1891 the first group strove to rescue the peasants from
their poverty, while the second did the same for the workers. The
populist trend in the countryside inspired by Mikhailovsky and the
militants of 'Land and Liberty', like Breshko-Breshkovskaya and
Victor Chernov, formed a 'Socialist Revolutionary' party. At the
same time the urban trend captured by Marxism and inspired by the

'Liberation of Labour' group of Georgy Plekhanov, which was joined by Vera Zasulich, Axelrod, Martov, Lenin and Trotsky, became the 'Social Democratic' party. And the liberal trend in the strict sense of the word, the Union of Liberation, was organized as the 'Constitutional Democratic' party (known as Cadets), under the aegis of Paul Milyukov, Peter Struve and Nicholas Berdyaev.

The situation was, then, that at the moment when the conditions of political life in Russia were changing radically, the Tsar had announced that, for his part, he would change nothing in the order established by his ancestors. At this time, true, politics was still beyond his ken. After the coronation his attention was focused on his relations with his uncles, who were actual coadjutors of the empire, and he had also to organize his new life as Tsar, husband and father. In addition, he had to think about his relations with the other Powers, a sphere in which family relations of another sort were being formed.

The drama of the coronation had given rise to Nicholas's first great family quarrel, at the very moment when the new sovereign was enthroned. This quarrel set against each other the three brothers of Grand Duke Sergei – the Tsar's uncle who was then Governor of Moscow and who, in order to minimize a catastrophe for which he bore some responsibility, urged that no alteration be made in the programme of festivities – and four other Grand Dukes, the Mikhailoviches, who firmly advocated a contrary view. In the end, most of the festivities that were to have followed the coronation were cancelled, with the exception of the grand ball held that very evening at the French Embassy, a circumstance which Wilhelm II did not fail to note with resentment.

Rows and family quarrels were frequent at the Romanov court. The Grand Dukes, uncles of the new Tsar, had always been violent and turbulent men, but in Alexander III's time they had been obliged to control themselves. Now that young Nicholas, their gentle nephew with the eyes of a gazelle, was on the throne, nothing held them back. They could change their wives or grab whatever woman they wanted. During the weeks that followed the coronation Nicholas obtained the divorce of Anastasia of Leuchtenberg; Michael,

Nicholas's brother, stole the wife of an officer in his regiment; Uncle Paul was able to take in hand the future Countess Paley; and Grand Duke Cyril married the Tsaritsa's sister-in-law.

The very week of the coronation, while Prince Yusupov, one of the Mikhailoviches, was at table with his father, the sound of a horse was heard from the next room. Suddenly, the door was flung open and there appeared a horseman with a fine presence, Prince G. Wittgenstein, one of the officers of the Tsar's escort, an attractive man on whom all the court ladies doted. He held a bouquet of roses which he cast at the feet of the Princess. Then he jumped out of the open window and disappeared.

Alexandra, prudish and austere, could not stand these pranks and this aroma of scandal. Besides which, she did not get on with her mother-in-law, Maria Fyodorovna, the widow of Alexander III, who was still a sparkling society lady. As Nicholas had never dared to ask her for the crown jewels, the Dowager Empress had kept them. But Nicholas yielded to Alexandra on another point, by dismissing one or two of his uncles from their offices – though he restored them at once, out of family feeling. Only his mother's liaison with A. Baryatinsky, an aide-de-camp, upset him, and the scenes between them were a palace joke. His face would then become severe, as when he spoke of the behaviour of his uncles.

Nicholas II had three great-uncles living: Constantine, Nicholas and Michael. The first-named lived in retirement in the Crimea, with a ballerina; the second was Inspector-General of Cavalry; and the third, Sandro's father, presided over the State Council and was Inspector-General of Artillery.

His four uncles were much more to the fore. The eldest, Vladimir, was an artist, a patron of the ballet school, and would be Diaghilev's protector. With his wife, Maria Pavlovna of Mecklenburg, he gave the grandest receptions in Petersburg. Alexei, a sensualist rather than an artist, was a great connoisseur of French cuisine. He was also fond of the 'petites femmes' of Paris, whom he invited into high-class restaurants, which made him a pioneer in the history of the French dining-table, because at this time it was still the custom for men only to sit down together at banquets. Alexei was also High

Admiral of the Russian Navy. In reality, however, he was more interested in the past of this navy, its history, than in its future: he resigned after the sinking of his ships at Port Arthur. Uncle Sergei was the most active of them all. As Governor of Moscow he was the only one to engage in politics and try to find solutions to the crisis of the state. His reign in Moscow was marked by the tragedy at the coronation festivities, and culminated in his participation in the policy of Plehve, whose merits he praised to Nicholas II; both men were murdered by terrorists, one in 1903, the other in 1905. Uncle Paul was an excellent dancer, and also a philanderer, always incomparably well dressed: a dolman of dark green and silver, strawberry-coloured close-fitting breeches, the short riding-boots of the Grodno hussars. After his wife's death he committed the twofold mistake of marrying a divorcée who was not of noble birth. He was obliged to leave Russia, but came back during the war to command the Guards on the German front. He never involved himself in politics, but at the time of the revolution of February 1917 he was the only one of these men not to desert the imperial family but to try to save the dynasty.

Also around Nicholas were his cousins, the sons of his great-uncles. First, Constantine's sons, Constantine and Dmitri. The former, a poet when he felt like it, busied himself with the education of recruits. The latter was mainly a lover of horses. Then the sons of Nicholas: Grand Duke Nicholas Nikolaevich, whom the Tsar called Nikolasha, and Peter, who suffered from tuberculosis. Nikolasha wielded most influence over Nicholas II: he embodied the military order, led the Tsar's armies during the Great War and possessed an authority shared by no one else. The family were jealous of him and Alexandra hated him. His hand was seen, usually wrongly, behind all the misfortunes that descended upon the dynasty. He was blamed for the October Manifesto of 1905, whereas in fact he merely advised the Tsar to proclaim it so as the better to collaborate with the nation's elected representatives, and blamed also for his victories in 1914. In 1915, out of fear lest he become more popular than the Tsar, and urged by Alexandra to make this move, Nicholas II took his command from him. When the Tsar was imprisoned, in 1917, he blamed Nikolasha for having joined with the commanders of his armies in calling for his abdication, whereas the Grand Duke expressed that view only when everything had been settled.

The last branch, the Mikhailoviches, consisted, first, of Grand Duke Nicholas Mikhailovich, the intellectual and scholar of the family, a distinguished historian of Alexander I, more familiar with learned discussions at the Institut de France than with intrigues at the court of Petersburg. Secretly in favour of French-type parliamentarism, he was regarded at court as a visionary, but not a dangerous one. His brothers Sergei and George played no special role, hardly more than did another of Michael's sons, Alexander, known as Sandro – Nicholas's childhood friend – whose memoirs, *Once a Grand Duke*, throw a warm light on the personality of the Tsar. He kept his affection for the young man, and when his friend ascended the throne he tried, in vain, to open his eyes to the way the times were changing. Not that Sandro shared his brother Nicholas's ideas, but he would have preferred that 'the family' should not have been his Tsar's highest instance, with its laws and customs to be obeyed regardless of what policy enjoined. Whenever Alexander intervened, he was no longer, for Nicholas II, a Grand Duke like any other, or the old friend in whom he had confidence, but became once again the urchin with whom he used to play at Yalta.

Everything had to bow to the family hierarchy. The Tsar told Sandro: 'For nearly three hundred years it was a habit of my ancestors to insist upon military careers for their relatives. I am not going to break this tradition. I will not permit my uncles or my cousins to interfere with my government.'

Another of Nicholas II's principles was the separation of powers. Two incidents show how this dictated his conduct.

These uncles were powerful figures, peremptory in their manner, and the Tsar always looked upon them with fear; for him they were above all his father's brothers, and even after he came to the throne he continued to obey them. Their views were his orders, and they put all their weight behind them. Alexei, the admiral, weighed 250 pounds. They were tall, too: Nicholas, the general, stood 1.95 metres

(6 feet 4 inches) high. Sergei and Vladimir were less impressive, but just as demanding and authoritative.

Sandro, his dear Sandro, the inseparable friend of his childhood who had become the husband of the Tsar's sister Xenia, dreamt of reorganizing the navy, which he thought old-fashioned. But it was Uncle Alexei who ruled over the Tsar's ships of war.

'I am sure he won't like it. I tell you, Sandro, he won't tolerate it.'

'Great guns, Nicky, you are the Tsar. You can do whatever you feel is necessary for the protection of our national interests.'

'That sounds awfully good, Sandro, but I know Uncle Alexei. He will carry on terribly. Everybody in the palace will be certain to hear his voice.'

'I have no doubt about that, but so much the better. Then you will have an excellent excuse for dismissing him on the spot.'

'Fancy my dismissing Uncle Alexei! My father's favourite brother! Do you know, Sandro, I believe that my uncles are right and that you did turn socialist during your stay in America.' And it was Sandro who lost his job.

This conversation took place at the very beginning of the reign. But nothing had changed fifteen years later. On this occasion, in 1911, the episode involved Grand Duke Nicholas. The War Minister, Sukhomlinov, had prepared an immense *Kriegspiel*, just as General Kuropatkin had done in 1902. All the district commanders had been summoned to the Winter Palace, where the 'game' was to be played under the delighted chairmanship of the Tsar, who would allot roles. Half an hour before proceedings were due to begin, Grand Duke Nicholas, who hated Sukhomlinov, had a talk with the Tsar, and the whole thing was cancelled. The commanders were received one by one, for discussion of matters of routine, and then at once went back to their districts. The War Minister offered his resignation, but the Tsar refused to accept it.

Alexander III was fond of the palace at Gatchina, but his Empress, the gracious and sprightly Dagmar of Denmark who had become Maria Fyodorovna, preferred St Petersburg and its festivities. Rather than the Winter Palace, which she found too large, she had chosen

the Anichkov Palace, which was smaller and was situated at the end
of the Nevsky Prospekt. Nicholas II stayed there at the beginning
of his reign. It seemed to Alexandra that Maria Fyodorovna was
continuing her reign. Later, the new Tsar and Tsaritsa moved to
Peterhof, or Tsarskoe Selo, where they were able to live more freely,
à la russe, close to the countryside. Nicholas rose between 7 and 8
a.m., not making any noise, so as not to wake Alexandra. He said
a prayer, then took a bath. About 8.30 he breakfasted, drinking tea
with his wife. After that he received the aide-de-camp on duty and
the marshal-in-chief of the court, Count Benckendorff, who discussed
with him the ceremonies of the day. Then he received the commandant of the palace – successively Trepov, Dedulin and Voyeikov – the official who was most directly in contact with the Tsar,
because he was responsible for his security. Following this, Nicholas
received his ministers until 11.30. After taking a short walk with his
dogs, he shared a meal with his escort: cabbage soup or beetroot
soup, kasha and kvass. Then he went back to the palace for a second
series of receptions, and lunch in the rosewood room.

Alexandra herself composed the menus, the previous evening, with
each dish ordered in quantity sufficient for ten persons, so that the
Tsar or the Tsaritsa could invite whoever they liked without having
to warn the kitchen. Every meal comprised four courses – five in the
case of dinner, plus the hors-d'oeuvre. Nicholas often ordered, as an
hors-d'oeuvre for himself, sucking-pig with horseradish. He had
been obliged to give up the dish he liked best, fresh caviare, owing
to indigestion. He mostly drank port.

After lunch Nicholas received or worked until 3.30 and then went
for a walk until 5.30 precisely, which was tea-time. He profited by
this break to read the Russian newspapers of all tendencies, while
the Tsaritsa read the English ones. From 6 to 9, more audiences.
Dinner was always a highly formal occasion. After dinner, Nicholas
received the Prime Minister, who was preceded by the palace commandant. To end the day, the imperial couple went for a sleigh-ride,
guards along the route having been posted since 9. The sleigh, drawn
by two horses, was followed by a single horseman and then two
other carriages. For this occasion Nicholas wore the cap of the

imperial archers' regiment, which bore a cross. On their return at 11, their majesties took tea and Nicholas read aloud to his entire family. He liked doing that, and had a good knowledge of Russian literature. He read Turgenev's *A Sportsman's Sketchbook* or Mer-ezhkovsky's *Peter*. And when his daughters reached the age of eight or nine, he also read to them in French: *Le Bourgeois Gentilhomme*, or *Tartarin de Tarascon*.

At midnight a sentry took up position before the door. Nobody must talk or sneeze, so as not to wake the imperial couple. Before going to bed, Nicholas made the day's entry in his diary.

Except when he went hunting, Nicholas II thus busied himself with government matters for only two or three hours a day, his reading of the newspapers included. But what did he actually know? 'Living in the seclusion of Tsarskoe Selo,' Senator A. A. Polovtsev wrote in his diary in 1909,

one learns only in bits and snatches what is going on in the circles that guide the destiny of our country. There is no policy based on principle, well reasoned and firmly directed, in any field. Everything is being done piecemeal and fortuitously, on momentary impulse, through the intrigues of one person or another ... The young Tsar is becoming more and more contemptuous of the organs of his own authority and is beginning to believe in the beneficent effect of his autocratic power, which he exercises sporadically, without prior deliberation and without reference to the general course of affairs.

Everything was related to the proposals made by the Tsaritsa, the Grand Duchesses and all the intriguers with whom the court swarmed. If we consult the *Tabel o Rangakh* ('Table of Ranks') we find that in 1900 there were more than forty Grand Dukes and Grand Duchesses, not counting the hundreds of highly placed personages who made up the households of the Tsar and the Tsaritsa. A full list of them was given in the *Pamyatnaya Kniga*, a sort of *Who's Who* of the imperial court, which came out annually and was looked forward to with impatience and curiosity. A tiny book, with between 780 and 800 pages, printed on India paper, bound and embossed in gold, it also included a calendar of festivals, receptions and anniversaries, and the general rules of etiquette. This jewel of Russian publishing, with a quality of printing never equalled later, was, of course, the bedside book of members of the imperial court.

But from all sides Nicholas II heard that salvation was to be sought in autocracy. 'Great disasters are to be feared,' concluded Count Polovtsev.

'The Empress's extraordinary shyness is surprising,' said Secretary of State Taneyev, the father of Anna Vyrubova, who was the Empress's companion and loyal friend right down to her death, 'but she has a man's intelligence.' She was above all a mother, and it was with her first daughter, Olga, in her arms that she signed papers and reports. After Olga she was to carry Tatiana, Maria and Anastasia.

Anna Vyrubova tells us that sometimes, while Alexandra was reading a report, 'there would come a clear, musical whistle, like a bird call. It was the Emperor's special summons to his wife, and at the first sound her cheek would turn to rose, and, regardless of everything, she would fly to answer it.' It is Anna, too, who says that, when the Empress was young, her eyes 'wore an expression of constant merriment which explained her family nickname of "Sunny", a name, by the way, nearly always used by the Emperor'.

Life at court was at that time gay and carefree. In her first winter in Russia, Alexandra went to thirty-two balls, as well as engaging in other amusements. In 1903 the Tsaritsa was pregnant again. Would it, at last, be a boy? Anna Vyrubova writes:

The following summer the heir was born, amid the wildest rejoicing all over the empire ... The Emperor, in spite of the desperate sorrow brought upon him by a disastrous war, was quite mad with joy. His happiness and the mother's, however, was of short duration, for almost at once they learned that the poor child was afflicted with a dread disease, rather rare except in royal families, where it is only too common. The victims of this malady are known in medicine as haemophiliacs, or bleeders ... Although the boy's affliction was in no conceivable way her fault, [the Empress] dwelt morbidly on the fact that the disease is transmitted through the mother and that it was common in her family. One of her younger brothers suffered from it, also her uncle Leopold, Queen Victoria's youngest son, while all three sons of her sister, Princess Henry of Prussia, were similarly affected. One of the boys died young and the other two were lifelong invalids.

The Emperor and Empress guarded their secret from all except relatives and most intimate friends, closing their eyes and their ears to the growing unpopularity of the Empress. She was ill and she was suffering, but to the court she appeared merely cold, haughty and indifferent. From this false impression she never fully recovered, even after the explanation of her suddenly acquired silence and melancholy became known.

Alexandra realized very well that it was necessary to re-establish her contact with her husband's subjects. But, like him, she blamed the alienation upon the intellectuals. To Prince Svyatopolk-Mirsky, the Minister of the Interior, who told her at the beginning of the reign that 'everyone in Russia was against the existing order', she replied: 'Yes, the intellectuals are against the Tsar and his government, but the people as a whole have always been for him and always will be.' She convinced herself that this was so, because she suffered from a confused sense of guilt that obliged her to believe it. A German, she wanted to be as Russian as the Tsar; a Protestant, she had changed her religion in order to marry him; a mother, she had produced four daughters before giving him an heir; and from the boy's birth he was known to be doomed, on account of *her* blood.

She thereafter took refuge in ceremonial and mysticism, and Nicholas too felt at home there. Never had the St Petersburg 'season' been so prestigious as at the beginning of his reign. And never, over against the atheist élite, had the court seemed so remote, so preoccupied with God.

When Alexei's illness was found to be incurable, Alexandra herself fell ill. Her health was delicate, and sciatica and successive pregnancies had already worn her out. She had attacks of vertigo and her constant state of anxiety regarding Alexei undermined her constitution. Furthermore, as she grew more and more neurasthenic, only her faith and her sense of duty came to matter to her. She was required continually to appear in public, which she hated doing.

Distressed and irritated, Nicholas wrote to his mother: 'She spends most of her time in bed, lives in seclusion, doesn't come to meals and looks out of the window for hours together.'

But it was the secret, the dreadful secret, that was wasting her. At any cost it must be kept from the people that the Tsar would not be able to have a proper successor. And the concealment of the truth, the wall of silence, contributed to isolating the imperial couple still further from their people.

Cruises on the imperial yacht, the *Standart*, gave relaxation, and also visits to Livadia. The wonderful climate of the Crimea, the scent of flowers and fruit and their brilliant colours helped greatly. The peninsula had kept its beauty intact because Alexander III and, after him, Nicholas II had banned the introduction of railways and other modern things that spoil landscapes. Nicholas had caused to be built there a palace of stone and marble, shining amid the greenery, and his family would not have missed for anything their Easter holiday in this palace.

Every year, after the church service, the festivities began and they ate *paskha* and then the *kulich*, the round cake crowned with the symbol 'XB', the Russian initials of '*Khristos voskres*', 'Christ is risen'. The Tsar and Alexandra bestowed on all their guests the triple kiss required by ritual: blessing, welcome, good wishes. Then, following a tradition that dated from Alexander III's time, Peter Carl Fabergé, the famous jeweller to the Romanovs, presented two of his creations, one for the Tsaritsa, the other for the Dowager Empress – eggs encrusted with precious stones and diamonds. Fifty-six were created for these occasions between 1884 and the end of the regime.

During the good years the season began with the charity sale organized by Grand Duchess Maria Pavlovna, widow of Grand Duke Vladimir, which went on for four days in the Hall of the Nobility. Then came the series of balls. Between Christmas and Lent every week was occupied. The girls and young married women distinguished themselves by rarely attending the same balls and not sharing the same escorts.

For the girls on their own a white ball was given. Few outsiders were invited, and whole rows of chaperons lined the walls, watching to see that no girl danced more than twice with the same partner. A girl had to curtsy to each of these old ladies before she began to dance, and if she was not known to them already, she had to be formally presented. They were all dressed in black, grey or purple satin, had fur stoles round their shoulders, wore splendid pearls, and their hair was combed back and smooth. These white balls had one advantage: the master of ceremonies saw to it that no girl lacked a partner.

There were never waltzes at a white ball and the two-step was considered to be not quite proper. Nicholas II had forbidden any officer in uniform to dance the tango or the one-step. Most of the evening was spent in quadrilles, with the master of ceremonies calling out his instructions: 'Advance, retire, take hands, form a circle ... ' Starting formally, quadrilles ended in wild disorder.

Then came the turn of the cotillion, the chief attraction of the ball, which was the subject of endless negotiations conducted by the master of ceremonies in order to match the couples. The figures did not differ greatly from those of the quadrille, but the dancers offered each other flowers, gifts, balloons of every colour, rosettes made of ribbon, and little bells. High hedges of huge pink roses were brought in, over which girls reached on tiptoe to grasp the hand of some unseen partner, hidden under heaps of jonquils and carnations brought in wagon-loads from the south of France, as we learn from Meriel Buchanan. At last it was time for the grand supper, to which each girl sat down with her cavalier. The little tables were always gayer than the big ones, occupied by the chaperons.

At the other balls things were less conventional. The orchestra played one-steps, and sometimes Gulescu's gypsy orchestra would change the ambience completely, and the dancing went on till morning.

Since there would be no more dancing once Lent had begun, the last week of carnival-time was devoted to balls. The entertainment was in full swing on Sunday, as the fast began at midnight. That crazy day everyone went from one festivity or ball to another, varied by sleigh-rides or trips on toboggans. This was certainly a very tiring way of life, but the buffets laden with *zakuski* enabled one to regain vigour. Waiting to be eaten were little raw mushrooms with cream, hot sausages cooked in wine, three kinds of caviare, little salted cucumbers and steaming shchi or borshch (cabbage soup or beetroot soup). Between each serving there were small glasses of vodka to be drunk, or yellow Russian cigarettes to be quickly smoked.

Young Meriel, the daughter of the British Ambassador, complained laughingly of this pomp and protocol, preferring, she said, to take a meal by the fireside. 'How foolish you are,' the French Ambassador said to her.

You don't know how many people envy you, and you cannot guess how some day you may look back on all this with a regret too deep for words ... Who knows, after all, how long it may last? Who knows what storms may come to sweep it all away? That Cossack, standing behind the chair of the Grand Duchess – what a gorgeous bit of colour, that scarlet soutane – he is a symbol of a vanishing tradition, and soon all that he stands for may be but a memory, and all this luxury and opulence which seems to us so secure and inviolate may vanish with him into obscurity.

The Tsar's extended family was made up of crowned heads. Cousin german to 'Willy' – Wilhelm II – Nicholas had married a granddaughter of Queen Victoria. The Romanovs, the Hohenzollerns and the Hessian and Hanoverian rulers, British or Germans, were closely connected and often found themselves assembled, for celebrations and engagement parties, as guests of Christian IX, the King of Denmark, who acted as a sort of head of the family. A daughter of his had married Alexander III, so that he was Nicholas II's grandfather. Another of his daughters had married the future Edward VII of England and a third the Prince of Hanover, while his son had been chosen to be King of the Hellenes. Family quarrels sometimes cut across conflicts between states, but these 'connections' could also facilitate inter-state relations.

Nicholas II kept family and political relations separate, and made this a matter of honour. Yet he enjoyed being with all these princes who assembled at Copenhagen in the summer. The bonds between generations, formed in childhood and renewed during these gatherings, left their mark, despite the chances of politics. It was no accident if on several occasions the court of Denmark mediated in conflicts between Britain, Germany and Russia. Nor was it an accident if, later, Willy and Georgie (Edward VII's son, the future George V) were to be more concerned than others about the fate of Nicky, his wife and his children.

Behind mainstream history there was thus formed a parallel history which sometimes intersected with it, without wishing to and without admitting that this was happening. This occurred on at least two occasions, in 1905 and 1918.

Wilhelm II tried to endue these family relations with a political legitimacy by urging Nicholas to see the bond between their two dynasties as contributing to the logic of alliances, whereas the Franco-Russian alliance, between a monarchy and a republic, was against nature. He developed this idea in a letter of 25 October 1895:

It is not the fact of the *rapport* or friendship between Russia and France that makes one uneasy. Every sovereign is sole master of his country's interests and he shapes his policy accordingly – but the danger which is brought to our principle of monarchism ... The constant appearance of princes, grand dukes, statesmen, generals in 'full fig' at reviews, burials, dinners and races ... makes republicans – as such – believe that they are quite honest, excellent people with whom princes can consort and feel at home!

... The republicans are revolutionists *de natura*, and where treated – rightly too – as people who must be shot or hanged, they tell our other loyal subjects: 'Oh, we are no dangerous bad men, look at France! ... The blood of Their Majesties [Louis XVI and his wife] is still on that country! Look at it, has it since then ever been happy or quiet again? Has it not staggered from bloodshed to bloodshed? Nicky, take my word for it, the curse of God has stricken that people for ever! ... I remember a few years ago a gentleman – no German – telling me full of horror that when he was at a fashionable *salon* in Paris he heard a Russian general answer a French one's question whether Russia would smash the German army: 'Oh, nous serons battus à plate couture, mais qu'est-ce que ça fait? Nous aurons aussi la république!' [Oh, we shall be beaten hollow, but what does that matter? Along with it we shall have a republic!'] That is what I am afraid of for you, my dear Nicky!'

The Willy–Nicky correspondence, published in 1924, covers the entire first part of the reign (1895–1908). We see from it that proposals were always initiated by Wilhelm II, while Nicholas, the younger man, said little or nothing in reply. Wilhelm annoyed him. Totally without tact, the German Emperor always sent his good wishes to Alexandra, a way of reminding Nicholas that he had helped to persuade her to marry him. Besides which, Nicholas realized quite well what his cousin's ulterior motive was, namely, to detach him from the alliance with France. He did not like the attitude of this older cousin fond of giving him advice, this man whom his father had jokingly called 'the dervish' on account of his way of dressing, which was always excessive and a little absurd.

'There are two imbeciles in history,' Alexander III used to say; 'Jan Sobieski, King of Poland, and my grandfather Nicholas I, because they both saved the House of Austria.' He was referring to the victory won by Sobieski in 1683, when the Turks were besieging Vienna, and to the intervention by the Russians in 1849, when they came to the aid of Franz-Josef in crushing the Hungarians, who had revolted in order to win independence.

When Nicholas II became Tsar in 1894 the House of Austria was still seen as Russia's enemy, with rancour and contempt. Not forgotten was Austria's 'ingratitude' during the Crimean War, when, far from returning the favour Nicholas I had done him in 1849, Franz-Josef had maintained a rather friendly neutrality towards Russia's foes, the British, French and Turks. Throughout the Balkan crises of 1878–87 the Tsar had always found the Habsburgs in the opposite camp.

Another neighbour viewed with suspicion and disdain was the Coburg prince who reigned over Bulgaria. On the pretext that his country owed its independence to Russia he showed himself touchy, haughty and even offensive in his dealings with the Romanovs, just in case they should try to take control of him. Paradoxically, in 1895

Russia no longer had diplomatic relations with Bulgaria, which owed its existence to her.

Neither the Habsburgs nor the Coburgs took part in the family festivities at Darmstadt or Copenhagen. The British, on the other hand, always turned up, and the Russians as well, even though the Tsars and their family regarded Britain as Russia's main enemy. Russia was, for Britain, 'that huge glacier moving down towards India'. 'One day India will be ours,' Nicholas wrote to his father during his voyage to the Far East, and Alexander III annotated this letter with the comment: 'Y penser toujours, n'en parler jamais' ('Think of it ever, speak of it never'). Besides, the Russians were at odds with Britain from one end of Asia to the other – over control of the Straits, of Persia, of Afghanistan, of Tibet.

There was a kind of rivalry between these two countries, the bear and the whale – one dominant on land, the other on the sea. When the princes and officers of these two nations met on Wilhelm II's yacht or at Hanover, they did not fail to challenge each other. There was even once an oratorical tournament at which, after responding to a score of toasts, one of the champions of Her Britannic Majesty, Colonel More-Molyneux, defended Victoria's cause against the Tsar's hussars. The Germans, for their part, had been eliminated from this tournament because they had limited the toasts to the persons present at the banquet, whereas the Russians had gone on drinking in honour of each prince, each regiment, and so on. More-Molyneux alone had 'stood firm' and had not been eliminated by the Russians.

In short, they respected the British.

The Russians did respect the Germans, too, but they had a traditional mistrust of Berlin which was not wholly due to the 'protective' air adopted by Wilhelm in relation to Nicholas II. The Tsar's mother, a member of the Danish royal house, hated the Prussians, who had seized Schleswig-Holstein in 1864, the year of her marriage. Alexander III's chief grudge against the Germans was that they had organized the Congress of Berlin, somewhat behind his back, in order to settle the Bulgarian question. In the days of the Three Emperors' Alliance the Tsar had felt caught in a trap because he was bound hand and foot in his relations with Austria-Hungary. Bismarck had agreed to conclude a reinsurance treaty with Russia alone, which implied that Germany would cease to give support to the anti-Russian zeal of the Austrians, but the treaty had not been renewed when Bismarck left office in 1890. This meant that confidence no longer prevailed between Berlin and St Petersburg.

The relative isolation of Russia in 1894 was such that Tsar Alexander III could say, when proposing a toast to King Nicholas of Montenegro:[4] 'You are my only real friend.' And that was almost true.

The alliance with France departed from the traditional framework of dynastic groupings and when Alexander III, in 1891, associated himself with the French Republic, this did not fail to arouse a hue and cry. From Russia's standpoint the alliance was dictated by necessity. Financial aid was needed from France, in the first place, a matter about which there was plenty of gossip, and there was also the need to avoid being paralysed by the alliances created by Berlin and Vienna, and to have a counterweight to Britain on the seas. Last but not least, a reason never mentioned by the French: Russia feared that the growing strength of the German Empire would result in France disappearing, as a nation, so that she had to be protected. *And France had to be protected, above all, from her own suicidal desire for 'revanche'.* This was one of the secrets of the Franco-Russian alliance as seen by Russia, and Nicholas II's conception of it. However, these purposes were not all to be revealed simultaneously.

On the morrow of Nicholas II's visit to France, Europe made fun ... of France. Caricaturists in Austria, Germany and Italy had a field day. *Jugend*, the Munich satirical paper, showed Madame Prudence, an elderly governess, giving Mademoiselle France advice to show reserve, but the crazy creature was ignoring this. She was throwing her arms round the neck of the Cossack Tsar, showing her legs and revealing all her charms. Elsewhere, the French cockerel, perched on the Russian bear, was flouting the Prussian fox, which was gazing

4. Montenegro remained Russia's ally right down to the war of 1914. In 1918 it became one of the constituent parts of the new state of Yugoslavia.

greedily at this prey that had gone out of his reach. *Asino*, a satirical paper published in Turin, was cruel. A heavy crown loaded with skulls was crushing the Russians deported to Siberia, who were sending to the French a symbolic reminder of their republican sentiments. 'It's the world turned upside-down,' they said in Vienna. 'Yesterday it was Marianne who tamed the beasts, but today it's the Russian bear that has her on a leash.' 'What is the difference between your empire and our republic?' asks a sansculotte. And the muzhik answers: 'Why, not much – you civilize the world with an axe and we do it with rope.' But the severest satire came from *Der Floh*. In its pages Marianne is a tart. In a restaurant she has plied the Tsar with drink, and now, standing before his empty seat, the waiter is saying to her, in allusion to France's loans to Russia: 'The gentleman has gone, leaving the bill for you to pay.' A prostitute swooning before her prince charming; it was not the Tsar's image that came weakened out of this visit, but that of the Republic. Nicholas II, on the contrary, appeared in the role of a knight, a protecting Caesar, while Félix Faure played the flunkey, ready to do anything to ingratiate himself with the Tsar.

Berlin's welcome, soon afterwards, was not so enthusiastic. It served no less, however, to convince the Tsar that his people did not appreciate their good luck in having a ruler like him.

Nicholas II was his own Foreign Minister, and his natural silence and reserve accordingly assured some secrecy for his plans. Though he never made them explicit, it is possible to detect their purposes, even though he surrounded these plans with a fog the reason for which did not become clear till later. Nicholas wanted the alliance with France; but, even more important than helping the country's economic development ('France is our money-box,' he was often to say), his purpose was, above all, to neutralize Britain while Russia carried out her 'grand design' in Asia and in the Pacific. This schema emerged in 1897, when Franz-Josef, Wilhelm II and Félix Faure all came to St Petersburg.

To Félix Faure, hailed by the Russians to the tune of the 'Marseillaise', Nicholas II repeated that the military alliance with France served only peaceful aims; he said nothing about France's *revanche* or Alsace-Lorraine. Even with the implications of the alliance restricted in this way the Russians considered that France gained by it, since it guaranteed her security in face of a threat the seriousness of which she doubtless failed to appreciate. But Nicholas II had no intention of entering into conflict with Wilhelm II.

The function of the alliance with France emerged still more clearly when the Tsar negotiated with Franz-Josef. It looked as though no agreement between them was possible; but, since the policy of Alexander II and Alexander III had resulted only in disappointment in the Balkans, Nicholas proposed to Franz-Josef to 'freeze' their dispute. The Austrian Emperor agreed, and this decision ensured peace in the region for more than ten years.

The British were better placed than other nations, especially the French, who were embroiled in a nationalistic crisis, to understand the implicit meaning of Nicholas II's actions. Did not the Russian's desire to pacify Franco-German relations and 'freeze' his quarrel with Austria mean that the Tsar wanted Europe to stay quiet so that he might be free to act elsewhere? Against the Turks? Or in the Far East? In either case, Britain was worried.

Two events showed clearly to the British where Russia thought her interests lay. First, Tsarist diplomacy intervened to mediate in the Greco-Turkish conflict, helping to achieve a settlement which, though favourable to Greece (she acquired Crete), nevertheless set a check to that country's ambitions. The conclusion drawn from this was that the Tsar had no intentions in *that* direction, where the British would have reacted immediately.

Second, in February 1898 the Tsar launched a gigantic naval programme costing 90 million roubles, the ships being destined almost wholly for the Baltic, the White Sea and the Pacific, but not for the Black Sea. Wilhelm II was, of course, worried by the Baltic aspect of the programme, and made a move towards a *rapprochement* with Britain. In fact, however, the ultimate destination of part of that new fleet was the Far East, which was the real direction of Nicholas's ambitions.

Nicholas's grand idea was disarmament and world peace. He had been impressed by his reading of Ivan Bloch's book *The War of the*

Future (1898), which showed the ravages that would ensue from a world war – seaborne trade ruined, Russia unable to export her grain, and so on. His ministers, especially Witte, had good reason to encourage the Tsar in this line of thinking. In 1898 Russia had just seized Port Arthur and so was replete; and, in Europe, the alliance with France had been consolidated. The purpose of that alliance had not been what the French wrongly supposed, namely, to help them recover Alsace-Lorraine, but rather to check France's inclination to *revanche* and secure her against possible attack by Germany. Its essential purpose was thus to prevent another war.

In this way Russia would be made free to do as she pleased in the Far East, it being understood that, from Nicholas's point of view, penetrating China, or 'coming up against the Japanese', did not mean going to war, any more than fighting in Afghanistan. Was the Tsar also afraid that the United States, having just overcome Spain in Cuba, would now get increasingly active in the Far East, and might ally with the British, or even the Japanese, so as to prevent Russia from accomplishing her 'mission'? This was a hypothesis held by Russia's foreign ministry, one that argued in favour of general disarmament and a reduction in expenditure for war purposes (even though Russia's debt had already, according to Oldenburg, fallen from 258 to 158 million roubles between 1897 and 1900).

All of these considerations applied, but, above all, the Tsar was peaceful by nature, at least where his relations with the courts and Powers of Europe were concerned. He believed in his mission in Asia and wished for peace to prevail elsewhere, his well-understood interest thus coinciding with his ideal. Expansion in China or other parts of Asia was, in his eyes, not war but a crusade for civilization, and it required peace in the West.

The Tsar put his personal signature to the Note of 16/28 August 1898 addressed to the Powers and composed in accordance with his views. He had overseen its formulation with the care of an attentive father. The Note was a call for disarmament. The response it met with was unanimously negative. In France people saw in it the hand of Wilhelm II, aimed at countering their efforts at military recovery since 1871. In Germany, on the contrary, Wilhelm suspected a trick designed to restrain the irresistible rise in the Reich's military power, and he sneered at Nicholas, who was doubtless unable to 'keep up' in the arms race. Britain, true, welcomed the Note – but added that its proposals could have no bearing on the Home Fleet. So that it was only Italy and Austria-Hungary who responded positively to the Tsar's initiative.

Although Generals Muraviev and Kuropatkin, who carried on the negotiations, explained that the Powers were not being asked actually to disarm, but only to restrict future increases in their armaments, this had no effect. The project was stillborn. It was also made clear that, so long as France had not recovered Alsace-Lorraine, peace and disarmament were out of the question in that part of Europe.

St Petersburg now produced the idea that Germany should restore France's lost provinces in exchange for Russia's backing in the creation of a Greater Germany, meaning inclusion of Austria in the Reich, through a dismantling of the Dual Monarchy. In return for Russian co-operation in bringing this about, Germany would, it was hoped, let the Tsar take over the Straits – as well, of course, as being allowed a free hand in the Far East.

The conference at The Hague proposed by the Tsar put an end to these senseless dreams. The conference took place, in spite of everything, since it seemed inopportune to offend the new monarch. Russia's proposals for disarmament programmed over a period of years were presented, but only as a formality. All the same, some decisions were taken with a view to reducing so far as possible the cruelty of future wars, which themselves were seen as unavoidable. At this conference, known as the First Conference for Peace, which was attended by over thirty states, including the USA, China, Mexico and all the European countries, the governments represented made several solemn promises: not to use asphyxiating gases (which, nevertheless, the Germans, French and British did use from 1915 onwards), not to use gas-filled shells (which the same armies fired in 1916) and not to use explosive ('dum-dum') bullets (which the Italians used in Ethiopia in 1934).

Actually, the first 'grand idea of the reign' had attracted nobody. Above all, it had failed to interest public opinion, which was hostile on principle, especially in Russia itself, to anything that came from an autocratic sovereign.

Nicholas's 'peace' policy had the aim of giving him freedom of action in the Far East, where his action would, he knew, be encouraged in any case by Wilhelm II. In his diary, General Kuropatkin, Minister of War and later commander-in-chief, records the dreams that the Tsar entertained at this time: to take possession of Manchuria, Korea and Tibet, and then to move on to Persia and the Straits. The Tsar thinks, Kuropatkin wrote, that we, his ministers, are hindering him in the realization of his dreams, out of petty considerations.

He persists in thinking he is right and understands better than we do what is needed for the glory and prosperity of Russia. Consequently, any Bezobrazov[5] who sings in tune with him seems to the sovereign to understand his intentions better than we, his ministers, understand them. And so he dissembles with us: but his experience and wisdom are growing rapidly and it is my opinion that, despite his natural diffidence, he will soon throw off mentors like Khlopov, Meshchersky and Bezobrazov, and directly and firmly impose upon us his views and his will.

The Tsar's will was to realize this dream: before entering as conqueror the city of Constantinople (the 'Second Rome', which he always called Tsargrad), to become Admiral of the Pacific. A group favouring this took shape around him, made up of Grand Duke Alexander Mikhailovich, Bezobrazov, Rear-Admiral Abaza and Admiral Alexeyev. Plehve, who was to become Minister of the Interior, approved of their plans. He said that, in order to put an end to the domestic malaise and the revolutionary mood that was beginning to arise, nothing would do more good than 'a little victorious war'.

During the coronation Li Hung-chang had made an agreement with Russia whereby, in exchange for his support, the Tsar would

5. A. Bezobrazov was a businessman close to Nicholas II who advocated Russia's expansion in the Far East.

be allowed to continue the building of the Trans-Siberian Railway through Manchuria. At this time the Chinese Emperor needed Russia's help, because Japan had infiltrated Korea and was encouraging there the reform party which aimed to overthrow the King, a protégé of China's.

War broke out between China and Japan in 1895, and demonstrated Japan's superiority on land and sea alike. Within a few months, after occupying Formosa (Taiwan), the Japanese were threatening Peking from the north, and China was forced to sign the peace treaty of Shimonoseki.

Russia had not been able to react in a military way, as the Trans-Siberian had not yet been completed; but the Minister of Foreign Affairs, Lobanov, convinced Nicholas that he must acquire a warm-water port on the coast of Manchuria and connected with Siberia. Helped by France, Germany and Britain, St Petersburg now brought pressure to bear on Japan that was strong enough to make the Mikado cede to Russia the Peninsula of Liaotung, where the Port Arthur base was situated. Nevertheless, Japan remained in possession of Formosa, and Korea became independent. What had been a setback for China, and for her ally, Russia, turned out to be a success for the Tsar, whose flag now flew over Port Arthur.

In reality, Russia's position was relatively less strong after the annexation of Liaotung than before, because the European Powers were now involving themselves more and more in Far Eastern affairs. The 'myth' of the Chinese market fascinated them all, and the virtual partition of the country into spheres of influence, the so-called 'break-up of China', became the principal subject of negotiations. Russia was not alone in getting directly involved in the question of China's future, since, on the margins, the British or the French might succeed in obtaining free ports.

The appearance of the Germans on this scene increased the scope of European intervention. Nicholas had agreed that Wilhelm should have a 'coaling station' at Kiaochow, where his ships could refuel, just as Russia used Port Arthur for this purpose. When, in 1897, two German clergymen were murdered at Kiaochow by the Boxers, Wilhelm wanted to send a punitive expedition. 'I can express neither approval nor disapproval of your orders regarding the dispatch of the German squadron to Kiaochow,' the Tsar replied, to a request that, in a way, recognized Russia's special position in the Far East. He added: 'I fear that severe measures taken to punish the Chinese may ... perhaps widen the abyss separating Christians from Chinese.' At the same time, he advised Peking to punish the guilty persons severely, so as to render pointless the sending of the punitive expedition.

That expedition took place all the same, and there ensued the 'Boxer campaign', into which Wilhelm was able to draw the other Powers, Russia included. Naturally, the Tsar followed Wilhelm so as the better to be able to hold him back. After this little war, Russia, like the other Powers, withdrew her fleet; but she kept her troops in Manchuria. For the second time, a foreign incursion into China had resulted in a gain for Russia. 'A state of affairs like this', observed France's Ambassador in St Petersburg, 'looks very much like a protectorate.' And the French naval attaché in Tokyo wrote (10 March 1901): 'The example of Egypt is there to enable us to perceive, if we don't let ourselves be put off by empty words, what is going to happen to Manchuria.'

When, in 1902, with the signing of the Anglo-Japanese alliance, the Tsar realized that he would have to retreat, and promised a gradual evacuation of Manchuria, the possibility of war between Russia and Japan emerged.

Sergei Witte, who had been in charge of Russia's finances since Alexander III's time, considered that a war would be a disaster. The Foreign Minister, Lamsdorf, warned the Tsar against an adventurous policy. Under their influence, the first zone of occupation in Manchuria was evacuated. However, Nicholas sympathized with the advocates of intervention – Bezobrazov, who aimed to exploit the country, Grand Duke Alexander, the Tsar's brother-in-law, Admiral Alexeyev, and also the War Minister, General Kuropatkin (who, nevertheless, did not want war). Russia's minister in Tokyo, Rosen, explained to his French colleague that his country could not go back on all the effort that had been put in for many years, and that was Nicholas's view as well. He took away the conduct of Far Eastern affairs from his ministers and entrusted it to Admiral Alexeyev, whom he appointed Viceroy directly responsible to the Tsar.

It is not surprising that Japan reacted to this step. The Japanese had armed quietly, and held the strategic advantage, and they had reason to fear that the Russians' naval plan might change the situation in their disfavour. Time was working against them.

In January 1904 Japan asked the Tsar to say whether or not he respected the integrity of Chinese territory in Manchuria. 'In the absence of a satisfactory reply', on 8 February 1904, Japan made a surprise attack, without any declaration of war, upon the Russian fleet stationed off Port Arthur, and destroyed part of it. Having thus obtained mastery of the seas, Japan landed her troops in Manchuria and forced the Russians to retire to Mukden.

The Russians suffered reverse after reverse. They had not, however, been defeated, and it was decided to send the Baltic fleet to the aid of what was left of the fleet immobilized in Port Arthur, a voyage that must take several months. Plehve's 'little war' had begun badly. The Tsar's diary records, day by day, week by week, the news, mainly bad, that arrived from the Far East. As he saw it, whatever was happening was God's will, and a sort of mystical fatalism hovers over these notes of his.

In the film *Arsenal* (1929) Dovzhenko alternates tragic scenes with extracts from the Tsar's diary in which he writes: 'Fine weather today'. Actually, historians are wrong if they deduce from the vacuity of the Tsar's diary that he played no role and was indifferent to the fate of the empire. Nicholas was the master of his own foreign policy, and in 1903 he was able to withdraw Far Eastern problems from Count Lamsdorf's sphere of competence, and willingly to let Count Witte quit the Ministry of Finance, since they both tried to stop him from getting bogged down in Manchuria.

Nicholas detested the Japanese and had made Russian expansion in East Asia the grand idea of his reign, a crusade for civilization and Orthodoxy. Furthermore, he dreamt of becoming 'the Emperor of the Pacific'. Consequently, he was disposed to listen to those who urged him to act in that direction, whether or not economic advantages would be gained. The simple fact was that, although well informed in the matter of international relations, Nicholas, along with his military advisers, underestimated the strength of the Japanese; and, like his advisers, he believed that the Mikado was bluffing in his resistance to Russia. Japan would not dare to confront the Russian Empire. So he dared to take the decisive steps.

Internally, the problems presented themselves differently. The Tsar was not interested in the political debate. He had absorbed the views of the conservatives, and that explains sufficiently his attitude of avoidance or rejection. The larger grew the circle of advisers who urged him to modify the regime, to give the empire a constitution, the greater was his mistrust of them. Every succeeding event strengthened him in this attitude.

Thus, one of the effects of his visit to Paris was that, for the first time, the young Tsar really thought about these problems. What right had the Russian liberals to criticize their Tsar so much when the republican French had given him such an enthusiastic welcome? The experience reinforced his hostility towards the liberals, the socialists and all the intellectuals. Consequently, the incomprehension between the young Tsar and the élites of his country was now greater than ever.

Even his ministers fell under suspicion. When he inherited them, like Pobedonostsev, from his father, he considered that 'their time would soon be up'. And when he discovered that he could not do without them, that offended his self-esteem. The most obvious case was that of Witte, who was, moreover, opposed to his policy in Manchuria. Now, everybody admired Witte, the Tsar's mother especially, and this caused Nicholas profound annoyance. Witte had the sense of economy and, because of his competence, foreign bankers trusted him. He was responsible for the famous 'Russian loans' which brought so much money into his country. The railways, including the Trans-Siberian, were due to him, too, as was the development of large-scale industry and the first industrial fair held at Nizhny-Novgorod. This champion of industrialization was behind the steelworks of Krivoy-Rog, the factories of the Donbas and the opening up of the oilfields of Baku.

Though Witte was not, of course, the father of Russia's indu-

strialization, this technocrat *avant la lettre* was able, in the name of
liberalism, to find the means whereby to accelerate it, by gaining the
confidence of foreign banks, especially the French ones. He had the
idea of making the peasants pay for industrialization through indirect
taxes and establishing a state monopoly of vodka sales. He explained
his idea: either the muzhik works, in which case the grain surplus
will be exported (too bad if there are famines; we'll help the worst
off), or he won't work but will drink instead, and that will benefit
the state even more, as the sale of vodka brings money into the
Treasury. Between 1893 and 1899, 24 per cent of the government's
resources came from the vodka monopoly.

The peasantry were thus sacrificed to industrial progress. In the
poverty-stricken country, suffering kept on getting worse, and Russia
experienced some major famines during the first years of the reign.
Nicholas's concern for the victims is beyond doubt, but it was
confined to the organizing of relief.

Nicholas had to use the good services rendered him by Witte, but
he could not put up with him any longer and found him a burden.
He disliked him for being a Freemason and also for playing the role
of Cassandra with regard to the Far East, and being proved right.
The Tsar could not endure collaborators who made an impression
on their contemporaries. Witte was the first of them to suffer for this
fault, and later there would be Stolypin. During the Great War,
Nicholas dismissed Grand Duke Nicholas, who had too much influence on the soldiers and their officers. All these men offended him
because he had no magnetism and no particular competence, yet was
Tsar.

One day, before choosing a minister, Nicholas consulted old
Pobedonostsev, who replied: 'Plehve is a scoundrel; Sipyagin is a
fool.' Nicholas chose the fool.

The Tsar treated his ministers like servants. True, he did not
address them in the familiar manner, as his grandfather Alexander
II had done, but he did require of them a respectful attitude. Witte,
it was said, stood to attention before the Tsar, with his thumbs in
line with the seams of his trousers and bowing very low, almost
bending double, whereas elsewhere he stood up straight and had

nothing humble in his manner. V. Kokovtsov, who was for four years chairman of the Council of Ministers, left office without a farewell from the Tsar. 'In Russia ministers have no right to say what they really think,' S. Sazonov explained to his foreign colleagues. They owed everything to their Emperor, and in particular their salaries and other advantages, which varied from minister to minister. Three-quarters of the Tsar's highly placed servitors belonged to the nobility, but only a third of them possessed much property, so that, for most, personal fortune depended on their attitude to him.

Nicholas lacked confidence in his ministers and changed them more and more frequently. He had eleven Ministers of the Interior between 1905 and 1917, eight Ministers of Trade and nine Ministers of Agriculture. This movement speeded up during the 1914 war. In three years he had four chairmen of the Council of Ministers, four Procurators of the Holy Synod and six Ministers of the Interior. 'A real game of leap-frog,' said V. Purishkevich, one of the leaders of the Right.

The Tsar was trusting in his relations with the head of protocol, with the commander of his personal guard and with his children's teachers, but not with the high officials of the state. In so far as he favoured any of them, it was the traditionalists – Plehve rather than Witte. Yet the latter, 'the most intelligent man in the empire', did all he could to save him. He wanted to regenerate the country by means of the economy, and thereby to divert the anger of a discontented people. A conversation he had with his rival Plehve provides a good illustration of the two sets of ideas that were in conflict at that time, though both were concerned to defend the autocracy, for Witte was not a liberal in the political sense, even though he believed in modernizing Russian society. When, during the Revolution of 1905, he had to draft a constitution and someone asked him what he felt about this task, he replied: 'I have a constitution in my head, but as to my heart' – and he spat on the floor. He was not a traditionalist, and had said to his rival, Plehve, three years earlier, referring to the reforms of Alexander II: 'The building has been reconstructed, but the cupola has been left untouched.' Society was now demanding a share in law-making and the provision of some checks on the bureaucracy, and if its demands were not met, the resultant discontent would take illegal forms. Plehve replied: 'If we are incapable of changing the course of historical events which will shake the state, then we are obliged to dam the current, to hold it back, not to swim with it.' Reforms were best carried out by the government, which possessed experience. Should the opposition take power, 'there will come to the surface, led by the Jews, all the harmful, criminal elements that yearn for Russia's ruin and dissolution.' Nicholas II agreed with Plehve.

The Tsar's choice of ministers thus depended either on caprice or on circumstances. One day Grand Duke Sergei, his uncle, advised him to appoint Plehve, a strong man. Nicholas's mother preferred Witte. Soon Rasputin was to advise the appointment of ministers such as Protopopov. But what is important is that the Tsar chose them all one by one. He had no real Prime Minister – each minister was responsible directly to the Tsar and to him alone. Moreover, they were not allowed to resign, which would have meant that they had a personal opinion, a will of their own, in contradiction with the service they owed to the Tsar. The consequence was that, in the absence of 'cabinet solidarity', the ministers fought against each other and no government policy was ever developed. Expenditure was undertaken without informing the Minister of Finance, and the Minister of Justice did not know what his colleague the Minister of the Interior was up to. Consequently there could be no consistent direction in government, or, rather, several different directions were followed at the same time.

The Tsar was the sole master of what was put before him, and he concerned himself even with the smallest details. This was an effect of the ever-increasing centralization of power, which the border nationalities, especially Finland, tried to resist. In his picture of the Russian Empire at the end of the nineteenth century, Leroy-Beaulieu quotes from the *Official Messenger* and the *Bulletin of Laws*:

On the 15th of May, His Majesty deigned to grant his consent to the endowment, in the hospitals of Nizhny-Novgorod, of four beds for old men, from a fund of 6,300 roubles, bequeathed by Madame Catherine D., widow of the late General D. On the same day His Majesty vouchsafed his consent to the creation of: first, a scholarship to the First Gymnasium of Kazan, from a fund of 5,000 roubles, bequeathed by the widow of Court Councillor F., second, a scholarship to the boys' school of P., from a capital of 3,000 roubles, taken from the receipts of that locality.

It was as the outcome of innumerable requests that such decisions as these were taken, in the sphere of art, science or any other. Prior to the reforms of Alexander II, 'no public building – not so much as a village church or schoolhouse – could be put up without the plan being sent from Petersburg ... In the Emperor Nicholas I's time there were for every class of building three or four types or models, approved by the sovereign.'

This field of activity was the only one in which the centralizing of administration in Tsarist Russia had managed to go further than in republican France. Russia's bureaucracy, unlike that of France, did not have to face criticism from representatives of the nation. It was confronted only by poets.

Considering that he had not long to live, Tolstoy sent from Gaspra, in the Crimea, in 1902, a message to the Tsar which he saw as resembling a bottle cast into the sea, because he doubted that Nicholas II would ever read it, or, if he did, would take any notice of it. 'I should not like to die', wrote Tolstoy,

without having told you what I think of what you are now doing – told you what you could be doing, told you what great happiness your deeds could bring to thousands of beings and to yourself, but what great misery they may bring to them and to you if they continue to follow the same direction as at present ...

One third of Russia is under a regime of reinforced surveillance, that is to say, it has been outlawed. The army of policemen, regular and secret, grows continually. The prisons and places of deportation are filled with persons sentenced for political reasons, not to mention the hundreds of thousands of ordinary prisoners, to whom the workers must now be added. The censorship has attained a level of oppressiveness unknown even in the abominable period of the 1840s. Religious persecution has never been so frequent or so cruel, and grows worse every day. Troops with weapons loaded ready to fire on the people have been sent into every city ... There have already been cases of fratricidal bloodshed, and fresh ones, even more cruel, are being prepared. And the peasants, all one hundred million of them, are getting poorer every year, despite the expansion of the state budget, or else because of it. Famine has become a normal phenomenon. Normal likewise is the discontent of all classes of society with the government.

The reason for all this is obvious, namely, that your advisers assure you that by checking every vital movement among the people they are ensuring the people's prosperity and your own security.

But it would be easier to check the course of a river than the eternal forward march of mankind, ordered by God.'

Did the Tsar even learn of these warnings addressed to him by a man who had been excommunicated? He knew, at any rate, that the wonderful singers and performers who quickened his heartbeat were against him.

One evening at the beginning of the century the chorus of the Opera intended to present a petition to the Tsar. It was resolved that, after one of the first scenes in *Boris Godunov*, the curtain would rise to reveal the entire chorus in an attitude of supplication, looking towards the imperial box, while someone read out the petition.

When the curtain rose, according to plan, Chaliapin, who was not in the plot, was still on the stage. Thus, Tsar Boris was standing there, in front of the petitioners, dignified, colossal, a personification of imperial authority in his golden robes, with Monomakh's cap upon his head. For a moving moment Boris confronted Nicholas II. Then, instinctively, 'Boris' bent the knee, joining the petitioners, to the great fury of part of the public, young revolutionaries who would have preferred him to remain standing before Nicholas II, as though challenging him. But at whom were those boos directed?

In France, Rochefort had written a few years earlier, there were 35 million subjects, not counting the subjects of discontent. In Russia they numbered more than 100 million.

What did the Tsar know about them? He was at first aware of a sort of confused irritation. Six months after the catastrophe at his coronation there were demonstrations of protest against the indifference of the regime, which had still done nothing for the victims. There were almost 700 arrests, and in the universities, where the movement began, more than 200 students were sent down. This movement, though political, had no definite aim, but was all the more revealing for that. Next came the explosion of anger of the Poles and Finns. To the former, after a reign in which, as in sixteenth-century Spain, the formula 'one ruler, one faith, one law' had been applied, Nicholas II had at first said, during a visit to Warsaw, 'that he was ready to come to an understanding with the Poles provided that they stayed loyal to him'. In fact, he then proceeded to unveil in Wilno (Vilnius) a monument commemorating the repression of the revolt in 1863.

In Finland the governor, F. Seyn, introduced a regulation increasing the period of military service to five years, whereas an agreement with the Seim, the Finnish parliament, had defined what the country's obligations were when it was joined with Russia. Even the Finnish senate, consisting mainly of the Tsar's own appointees, refused to publish this decree, which, they said, 'shows contempt for the pledged word and for our constitutional laws'. This was followed by a petition with 500,000 signatures, an extraordinary number from a nation of fewer than 3 million. It appealed to the example of Alexander III, who had honoured the imperial pledge. Nicholas II considered that those who presented this petition were seeking to separate him from his people. He signed a second decree, introducing the Russian language into the administration of Finland as from 1905.

This policy of Russification, bitterly resented, provoked resistance even among the Armenians, the most loyal of the Tsar's non-Russian subjects. It was no accident that most of the 'national' organizations came into being in the first five years of Nicholas II's reign. Nothing was to be hoped for from the son of Alexander III.

There was a link between the Finnish revolt and the student riots which occurred soon afterwards. The students considered that they, too, were deprived of the elementary contractual rights that every civilized state ought to acknowledge for its citizens, and they demanded respect for the inviolability of their persons (a demand that tells us much about police tyranny) and publication of all the measures adopted that affected them.

To the students of St Petersburg the university rector's initial reply was that 'the birds of paradise, who are given whatever they ask for, do not flourish in our climate'. The demonstration that ensued brought out 2,500 students, and 25,000 workers went on strike in support. Nearly all the students were sent down from the university, which was closed. When it reopened, 2,181 students out of 2,425 were readmitted. The Minister of the Interior wanted to punish the excluded students by calling them up into the army, but General Kuropatkin refused to accept them. Nicholas II rebuked the students, saying: 'They should be studying, not putting forward demands. And their demands are not only pointless, they are harmful.' The Minister of Education, M. Bogolepov, was assassinated; and his killer, the Socialist Revolutionary P. Karpovich, was hailed by the students. Terrorism thus reappeared, 'without any real connection', opined General Vannovsky, Bogolepov's successor, 'with the issues involved in the students' demands ... But the conditions for an explosion existed, owing to the isolation of the students, who had no ties with their teachers.'

The new fact was the appearance of these students as a new social force. This was not the case in the West, where the students were, rather, the vanguard of social groups in conflict with authority. In Russia the situation was different. A good half of the students came from the poorer classes; in Moscow, for instance, of 4,017 registered students, 1,957 were sons of non-property-owners and 874 were scholarship holders. This shows that, however autocratic the regime might be, it was fostering a certain mixture of social groups, a process which led to the formation of a new force and produced a whole generation of revolutionaries between 1905 and 1917.

With the students, revolt took to the streets. True, as Witte said, Russia had the good luck not to possess, as yet, a real working class.

But the country was advancing with giant strides to the stage when such a class would appear. The state was itself stimulating industrial concentration. Already in 1901 nearly half of the workers were employed in enterprises with more than 500 workers each. Increases in productivity and in the size of factories went together, which showed that structural change, in St Petersburg as in Moscow, was tending towards 'giantism'.

Given the rate at which Russia's industrial equipment was growing in the first years of Nicholas II's reign, there can be no doubt, as the economist Gerschenkron wrote fifty years later, that, but for the Bolshevik Revolution, Russia would by now have overtaken the United States.

An admirable projection. Except that this irrefutable argument, thought up by a brilliant economist of the Cold War period, leaves out of account the fact that among the factors contributing to this economic progress were an eleven-hour working day and starvation wages. Consequently, revolution appeared, as an unwished-for companion of economic growth.

A grand idea of the reign which emerged during the student demonstrations was to separate political demands from the rest, and satisfy, at most, the non-political ones. This tactic, which enjoyed partial success in the universities, was subsequently applied to other fields of social life, and in particular to industrial conflicts. One of the high officials of the Ministry of the Interior, Sergei Zubatov, who was backed by Grand Duke Sergei, asked the professors of Moscow University who had participated in the negotiations of 1901 to draw up statutes for the university, and then to offer the workers statutes modelled on these. A good and convincing speaker, Zubatov had won the favour of Grand Duke Sergei, Governor of Moscow, with these words:

We have to win over the masses. They have faith in us, but the propaganda of the opposition and of the revolutionaries is aimed at reducing this faith. It is essential to revive it by showing concern: then, the opposition, however determined, will be helpless.

What this means is that the ideologues exploit the masses politically on the basis of their poverty, of their needs, and we need to take account of that fact.

By means of action at the root of these problems we disarm the masses through regular improvements in their conditions of life. The government must take charge of this work, on the basis of the masses' demands. Because this is all that the workers are asking for, at least for the moment. But the government must do this, and do it quickly and with consistency.

The principle of our internal policy must be to balance between the classes, which, at present, hate each other. An autocracy should keep above the classes and apply the principle of 'divide and rule'. Do not leave them time to agree between themselves: that would mean revolution, which we should help along if we were to take one side in this conflict, thereby confirming the arguments of the ideologists. We must create a counter-poison or antidote to the bourgeoisie, which is growing arrogant. Accordingly, we must bring the workers over to us, and so kill two birds with one stone: check the upsurge of the bourgeoisie and deprive the revolutionaries of their troops.

There soon came into being, officially constituted, under the aegis of Plehve and Zubatov, a 'Society for Mutual Help among Workers in Mechanical Production'. It was a kind of trade union, but under police protection. Patriotic demonstrations were organized in support of the Tsar, the workers met Plehve, and so on. The final episode in this scenario was the adoption of two new laws for the benefit of the workers. One provided for the election of 'worker seniors', chosen by their comrades to represent them in dealings with the employers, and the other provided for medical aid to victims of industrial injuries, half-pay during sick-leave and a grant towards workers' funerals. All this was intended to neutralize the action of the revolutionary parties that were being formed.

The Zubatov experience was extended from Moscow to Odessa and some other cities, but its initiator fell from grace through uttering some derogatory words concerning his boss, Plehve. Meanwhile, the professors, sharply criticized by the revolutionary intelligentsia, had retired into the background, and were replaced by priests. This did not always, however, have the same effect.

The Zubatovshchina had had an undesired dual consequence.[6] On the one hand, it antagonized the industrialists; and, on the other, it gave the working class a primary form of organization that the revolutionary groups proceeded to exploit. With the great strikes of 1903, which were partly due to the state-sponsored unions, the thing took such a bad turn in several Russian cities that the experiment was finally abandoned. But a movement had been started.

The idea was also to keep the workers' demands confined to their proper sphere, the enterprise, by creating an organizational form for them. Even if they were not enthusiasts for the Tsar, the workers would see him still as an arbiter. The workers' leisure was to be used to educate them, giving them ideas that would turn them against socialism.

In this context, repression had to be restricted to action against workers' activities undertaken *outside* of the enterprises where they were employed. Mass demonstrations such as those held on 1 May would be oriented in a direction favourable to the Tsar and to Orthodoxy. Going further yet, Grand Duke Sergei and Zubatov even helped to bring forward actual workers' leaders, taking care not to let this activity go astray – for example, by not allowing these men to play, even innocently, the role of double agent. Fyodor Slepov was one such leader. He was the very model of a conservative worker, anti-Semitic, very proud of himself. He attended the evening classes that were organized from above, just as the Social Democrats organized theirs. In these classes, instead of reading and discussing Marx or Zola, the programme included Tolstoy, labour law and the ideas of Paul de Rousiers, the father of trade union doctrine and the theory of relations between employers and workers, who was also an advocate of American-style democracy. The lecturers to these classes were, at first, university professors and jurists, but theologians and priests joined in the work, and one of these, Father Gapon, was soon to take over from Zubatov.

Naturally, the capitalists – and Witte – thought that all this was

6. In Russian the suffix *shchina* added to a name gives the concept a pejorative meaning ('the Zubatov business').

playing with fire. The workers' consciousness was being developed, and it did not matter if this was being done against socialism. For the moment, the capitalists considered that they alone were the sufferers from demands that, even though 'merely economic', nevertheless meant increased costs for them to bear. Witte thought that Grand Duke Sergei and Plehve were trying to oust him by causing him to lose the confidence of the capitalists. He was well aware that Nicholas preferred Plehve and Trepov to him because they were uncompromising in their political stand whereas he was disposed to agree to a constitutional compromise that would have benefited the bourgeoisie, something to which the Tsar was opposed.

Another perverse effect was that the Zubatovshchina provided the working class with a model of organization that the revolutionary leaders would exploit and that survived its 'initiator', namely, the soviet, a workers' council differing from the trade union in that its demands were also cultural and social. During the great strikes of 1903, which were due to some extent to these organizations created by a sort of cross-breeding between the state and the working class, the workers enjoyed such success in several cities that this caused alarm. The governor of Ufa fired on demonstrators at Zlatoust and there were violent clashes in Kiev, Nikolayev and other towns. In Odessa the strikes were so vigorous that the police ceased to tolerate them. When demonstrators called for 'a constituent assembly' the police opened fire, and Zubatov resigned.

The working class was no longer that politically amorphous mass described in *Iskra*, the organ of the Social Democrats. It was about to take the political stage, having been formed, in part, by the schools and the activities that the police had set going, parallel with the work of the Social Democrats.

Some months later a conflict at the Putilov works in St Petersburg repeated the Odessa scenario, but on an enormous scale and against the background of the defeats in Manchuria. Father Gapon, Zubatov's spiritual heir, organized a strike of 100,000 workers on 7 January 1905, and the numbers involved doubled next day. So astonishing and mysterious a phenomenon had never been seen before.

In the spirit of 'the grand idea' Gapon drew up a petition to the Tsar in which the workers appealed for his protection and for the just reforms they expected from him. Demands, fervour and faith came together in a document in which it was no longer possible to distinguish between service to the people, Orthodoxy, Holy Russia, love for the Tsar and the resurrection-revolution that would save society from socialism.

In three days the petition-appeal received more than 150,000 signatures. It mingled, pell-mell, arguments from the arsenals of liberalism, populism and Marxism, but the tone was that of Holy Russia. One hundred million muzhiks spoke with its voice.

Sire,

We, workers and inhabitants of the city of St Petersburg, belonging to different estates, our wives, our children and our parents, helpless old people, have come to you, Sire, to seek justice and protection. We have sunk into poverty, we are oppressed, we are crushed by an unbearable burden of work, we are insulted, we are not recognized as human beings, we are treated as slaves who must put up with their bitter lot and keep silent. And we have put up with it, but they are pushing us still further into the abyss of misery, lack of rights and ignorance. Despotism and arbitrariness choke us, and we cannot breathe. We are at the end of our strength, Sire. Our patience has reached its limit. That terrible moment has come for us when death is preferable to the continuance of unbearable suffering.

And so we have left our work and told our employers that we will not go back to it until they have agreed to our demands. We did not ask for much, we wanted only that without which life is not life but penal servitude, endless suffering. Our first request was that our employers discuss our needs with us. But they refused, they would not allow us the right to talk about our needs, because the law does not recognize such a right for us. Our requests also seemed to them to be illegal: reducing the hours of work to eight per day; drawing up a schedule of wage-rates for our work along with us and with our agreement; investigating our disputes with the lower management of the factories; increasing the wages of unskilled workers and women to one rouble per day; abolishing overtime; treating us with attention and without abuse; arranging the workshops so that we can work there without catching our deaths from frightful draughts, rain and snow.

To our employers and the managers of the factories all this seemed illegal, every one of our requests seemed a crime, and our desire to improve our position an impertinence insulting to them.

Sire, there are many thousands of us here, and we only look like human beings; in reality we, along with the Russian people as a whole, are not allowed any human right, not even the right to speak, think, assemble, discuss our needs, take steps to improve our situation. We have been enslaved, and enslaved with the help of your officials, under their protection and with their co-operation. Any one of us who dares to raise his voice in defence of the interests of the working class and the people is thrown into prison or sent into exile. Good-heartedness, a kindly spirit are punished as crimes. Showing pity for the downtrodden, rightless, suffering person means committing a grave crime. All the workers and peasants have been handed over to arbitrary rule by a government of officials made up of embezzlers and robbers, who are totally without concern for the interests of the people and who trample on those interests. This government of officials has brought the country to utter ruin, has drawn it into a shameful war and is leading Russia nearer and nearer to destruction. We, the workers and the people, have no say in how the huge sums extorted from us are spent. We do not even know where the money goes that is taken from the poverty-stricken people. The people have no power to express their wishes, their demands, to participate in determining what taxes should be levied and how the money should be spent. The workers have no opportunity to organize themselves in unions for defence of their interests.

Sire, is this in accordance with the divine laws by grace of which you reign? And can one live under such laws? Would it not be better for us to die, all of us working people throughout Russia? Let them live and enjoy life, the capitalists, the exploiters of the working class and the officials, embezzlers and robbers of the Russian people. That is the choice we face, Sire, and that is what has brought us together to the walls of your palace. Here we are seeking our last chance of salvation. Do not refuse to help your people, rescue them from the pit of rightlessness, poverty and ignorance, enable them to decide their own fate, free them from the intolerable oppression by your officials. Break down the wall between yourself and your people and let them rule the country along with you. Have you not been placed where you are for the well-being of the people? But the officials wrest this well-being from us, it does not reach us, we get only sorrow and humiliation. Consider our requests without anger and with care, they are not meant to do harm but to do good, to you as well as to us, Sire. It is not insolence that speaks in us, but awareness of the need to break out of a situation that is unbearable for everyone. Russia is too great, her needs are too diverse and too numerous for officials to be able to govern the country on their own. Representation of the people is necessary, the people must help themselves and govern themselves. After all, they alone know what their real needs are. Do not reject the people's help, accept it, cause to be convened, at once, without delay, representatives of the Russian land, from all classes and all estates, including representatives of the workers. Let there be here capitalist and worker, official, priest, doctor, teacher; let them all, whoever they are, elect their representatives. Let everyone be equal and free in their right to vote, and to that end decree that the elections to the constituent assembly be carried out under universal, secret and equal suffrage.

This is our chief request, all else is based in it and on it, this is the principal and only means of healing our painful wounds, without which these wounds will bleed heavily and bring us quickly to our death.

But one measure alone cannot heal our wounds. Others too are needed, and we will speak of them frankly and openly to you, Sire, as to a father, on behalf of all the workers of Russia.

What are needed are:

(I) *Measures against the ignorance and lack of rights of the Russian people*

1. Immediate release and return for all those who have suffered for their political and religious beliefs, for strikes and peasant disorders.
2. Immediate proclamation of liberty and inviolability of the person, freedom of speech, of the press, of assembly and of conscience in religious matters.
3. Universal and compulsory education at the state's expense.
4. Responsibility of the ministers to the people and guarantee of legality in administration.
5. Equality of all, without exception, before the law.

6. Separation of church and state.

(II) *Measures against the people's poverty*

1. Abolition of indirect taxes and their replacement by a direct and progressive income tax.
2. Abolition of the redemption dues, provision of cheap credit and gradual transfer of the land to the people.
3. Contracts for the army and the navy to be placed in Russia, not abroad.
4. An end to the war, in accordance with the people's will.

(III) *Measures against oppression of labour by capital*

1. Abolition of the factory inspectorate.
2. Establishment in the factories of standing commissions of workers' deputies, to examine along with the management all claims made by individual workers. No worker to be dismissed except by decision of this commission.
3. Freedom for workers' co-operatives and trade unions, at once.
4. An eight-hour day and regulation of overtime.
5. Freedom for labour to fight against capital, at once.
6. A basic wage, at once.
7. Obligatory participation by representatives of the workers in the preparation of a bill for state insurance of workers, at once.

There, Sire, you have our chief needs, which we have brought to you. Only by satisfying them can our fatherland be freed from slavery and poverty, and made prosperous, only thus can the workers organize to defend their interests against shameful exploitation by capitalists and by the bureaucrats' government which robs and chokes the people. Order these measures, and swear to carry them out, and you will make Russia happy and glorious; and thus you will engrave your name on our hearts and the hearts of our descendants, for ever. But if you do not order them and do not respond to our prayer we shall die, here, on this square, in front of your palace. We have nowhere further to go, and there would be no point in going. For us there are only two paths – either to freedom and happiness, or to the grave. Let our lives be a sacrifice for Russia, worn out with suffering. We do not grudge this sacrifice but offer it willingly.

Georgy Gapon, priest
Ivan Vasimov, worker

That day the Tsar was not in Moscow but at Tsarskoe Selo. Did he know what was happening? Certainly, because everyone knew that a grand appeal to the Tsar was being prepared. Even in far-off Paris this was known, where Grand Duke Paul, the Tsar's uncle, dining with Ambassador Paleologue, wondered why Nicholas was not receiving the strikers' delegates. And right down to a mere officer returning from Manchuria who learned in the train as it pulled into St Petersburg that 'tomorrow there is going to be a revolution'.

On Saturday 8 January, Nicholas wrote in his diary:

A bright, cold day. There was much work to do and a lot of reports for us to deal with. We had Freedericks[7] to lunch. I went for a long walk. Troops have been brought from the outskirts to reinforce the garrison. Up to now the workers have been calm. Their number is estimated at 120,000. At the head of their union is a kind of socialist priest named Gapon. Mirsky[8] came this evening to present his report on the measures taken.

A few days after the demonstration, at which the soldiers opened fire, killing 170 people – this was 'Bloody Sunday' – and wounding many more among the workers who, bearing icons and banners, peacefully approached their beloved Tsar, Nicholas addressed a carefully selected delegation:

By calling on you to come and present me with a petition concerning your needs, they provoked you to rebel against me and my government. Strikes and stormy meetings serve only to incite idle crowds to commit disorders such as have always compelled and always will compel the authorities to use military force, and that will inevitably result in innocent victims. I know that the workers' life is not an easy one. A great deal must be improved and put into good order. But have patience. You understand very well that it is necessary to be just towards your employers and to take account of the conditions in which our industry operates.

7. Count Freedericks, commander of the palace guard and Minister of the Court.
8. Prince Svyatopolk-Mirsky, the new Minister of the Interior. Plehve had been murdered a few weeks before this.

As for coming, in a rebellious mob, to tell me of your needs, that was a criminal act.

I have confidence in the honourable sentiments of the workers and in their unshakeable loyalty to me, and for that reason I forgive their offence.

In his diary the Tsar had written on 9 January:

A dreadful day. Serious disorders took place in Petersburg when the workers tried to get to the Winter Palace. The troops were forced to fire in several parts of the city and there are many killed and wounded. Lord, how painful and sad this is. Mama arrived from the city just in time for Mass. The family lunched together. I went for a walk with Michael. Mama stayed overnight.

On 10 January he wrote: 'Nothing special has happened in the city ... A deputation from the Ural Cossacks brought me some caviare ... I have decided to appoint General Trepov governor of the city and the province.' And on 19 January: 'I said a few words to a delegation of workers from St Petersburg, regarding the recent disorders.'

He forgave his people for rebelling against him. But his people did not forgive him.

2. The Autocracy versus Society: Nicholas 'the Bloody'

Bloody Sunday put an end to the myth of the Tsar-Batyushka, the 'Little Father' Tsar, the loving father from whom nothing but good could come. For a long time the people had thought that the nobles had kept them apart from their Tsar, turning him into someone who could no longer be approached or even spoken to. But the day was bound to arrive when Tsar and people would at last be able to understand one another ...

Yet when the people had come, with banners flying, to pray and speak to their beloved Tsar, his soldiers had replied with bullets.

Nevertheless, this faith in the Tsar had existed, and when the people revolted it was not against his person but against his henchmen. This had infuriated Herzen: 'You hate the big landowners and the bureaucrats, you fear them, and you are right to do so. But you have faith in the bishop, you believe the Tsar. Don't trust them: The Tsar is like the rest; they are *his* men.'

Until now the peasants had had good reason to draw the distinction. They knew that only the lands of Tsar Nicholas I had been reformed and some of his serfs emancipated, but that not a single nobleman had followed this example. They knew, too, that the great reform had been the work of Alexander II, the Tsar Liberator, that he had freed the peasants but that, subsequently, his agents had insisted that they pay for the land they worked. 'Society', the cultivated classes, those who in Russia were called *obshchestvo*, in contrast to the people, *narod*, spoke with irony about 'naïve monarchism'; but the muzhiks were not so naïve as all that. Their veneration and idolatry placed them, in a sense, under the Tsar's protection, and they could therefore act with impunity, 'in the name of the Tsar', against their landlords and against the nobles. This they were able to do, though, less and less as the nineteenth century progressed; for, with the help of the expanding railway network, the bureaucracy and the army tightened the mesh of centralization and paralysed their action. Bloody Sunday was an explosive revelation of this change in the situation.

After 1905 the peasants said no longer that they were acting 'in the name of the Tsar'. But the myth survived, in a way, since, when wearing soldiers' greatcoats, these same peasants had fired on the workers, who were peasants like themselves.

Gregor Alexinsky, a former Bolshevik who became a 'social patriot' in 1914, recalls words spoken by Maxim Gorky after 9 January 1905:

If I were Tsar of Russia I would act in such a way as to ensure the survival for ever of the absolute monarchy. I should go to Moscow and appear in front of the Kremlin on a white horse, with a substantial entourage, and I should say: 'Gather the people of Moscow around me here, at once!' When the people were assembled, I should say: 'My children, you are dissatisfied with my ministers, with my high officials, and with the rich men who plunder and maltreat you. Very well, I, your Tsar, will bring them to judgement on this very spot, those thieves and scoundrels, in your presence.' And after hearing the people's complaints, I should order some heads to be cut off there and then, without further judicial process. And I assure you that, after that, I should never again have to fear any attempt on my life. The people would protect me more effectively than any Okhrana.

So the old myth was still alive in a way, even for Gorky, though he was a sworn enemy of Tsarism. This writer came from the depths of the people and, despite his hatred, the idea of the Tsar-Batyushka was still rooted in him. But neither Nicholas II nor the circles hostile to the liberals were able to exploit this attachment to their advantage.

Bloody Sunday snapped the 'sacred bond' which had united the people with their Tsar – the people upon whom Nicholas founded his faith and the legitimacy of his rule. It marked also the irruption of the working class on to the historical scene: a turn of events that owed less to the revolutionary parties than to the action of the autocracy itself.

The Tsar's 'pardon' could not satisfy the people. A. Yermolov, the Minister of Agriculture, had the delicate task of explaining this to Nicholas: 'There is a risk that, one day, the soldiers may refuse to fire on unarmed people from whom they themselves come ... So it is necessary to appeal to the representatives of the Russian land ... to find support among the people as a whole.' A commission, known as the Shidlovsky Commission, was set up with the task of obtaining evidence as to the origin of the troubles. The workers agreed in principle to co-operate, despite the contrary view of many socialist leaders, provided that anyone they elected to participate in the commission's work was guaranteed immunity from prosecution. On 20 February this condition was declared unacceptable, and the Shidlovsky Commission was dissolved.

In the meantime Nicholas wrote in his diary: 'A fearful crime has been committed in Moscow. At the Nicholas Gate Uncle Sergei, who was travelling in a carriage, was killed by a bomb. The coachman was mortally wounded. Poor Ella [Sergei's wife and Alix's sister], may the Lord bless her and help her.'

The Tsar then ordered the Minister of the Interior, Bulygin, to draw up a 'rescript', a plan for the convoking of elected representatives. A stern manifesto calling on the people to rally round the Tsar was published in the press. The rescript, the manifesto and an *ukaz* inviting the population to make known their views and proposals all appeared together on 18 February. Though seeming to contradict one another, they nevertheless signified that, without modifying the autocratic principle, the Tsar was going to enter into a dialogue with the nation.

The idea of asking the views, but only the views, of the nation's representatives was an old one, having already been considered in 1881. However, the 'reaction' under Alexander III and then the policy defined by Nicholas at the beginning of his reign in his reply to the zemstvos had set it aside. The question nevertheless remained on the agenda, despite the monarch's resistance. In 1901, in the imperial proclamation marking the jubilee of the State Council,

Nicholas had struck out from the phrase referring to provincial conferences, 'in which men would participate who enjoy our confidence and that of the public', the words 'and that of the public'. Plehve had put them back after 1903, in a manifesto concerning improvement of the political system. This he did on the advice of V. Gurko, forcing himself to agree to those words – and the Tsar, for once, failed to react.

Contrary to legend, Nicholas II was very meticulous regarding definition of the rights and powers of which he made 'a problem of conscience'; and in February 1905 it was in full awareness, and subject to constraint, that he agreed to amend them.

After Bloody Sunday the rescript of 18 February looked like a retreat by the ruler. Excitement prevailed in the country. Students, professors, lawyers, industrialists (who, like the Jews, were used as scapegoats by the Tsarist authorities) and workers had multiplied strikes, demonstrations and petitions. Taking a stand, as requested by the Tsar's rescript, the population proceeded to organize themselves into associations of doctors, agronomists, lawyers, women, and so on, which soon came together in what was called the 'Union of Unions'. At the same time the workers also came together and set up, at Ivanovo-Voznesensk, the first 'soviet' known to history. This workers' council demanded for the workers 'the right to assemble freely to discuss their needs and elect delegates'. The socialist parties – Social Democrats and Socialist Revolutionaries – then only just coming into existence, had encouraged these movements (especially the Menshevik faction of the Social Democrats), which were all the more spectacular in that similar movements developed in Poland, where, since January, strikes had been widespread, orchestrated by the Polish Socialist Party and the Jewish Bund. From summer onwards the troubles began to affect the country districts, too.

The disaster of Tsushima (15/28 May), when Russia's Baltic Fleet, which had set out, via the Cape of Good Hope, to bring help to the Pacific Fleet, was destroyed by the Japanese after a voyage lasting seven and a half months, led to troubles in the navy, including the famous mutiny of the battleship *Potemkin*. The government immediately accused the rebels, both the mutinous sailors and the workers on strike, of being 'accomplices of the enemy', which was strictly true only of a legion of Polish volunteers who went to get military training in Japan. (It is true, also, that some revolutionaries who were more or less involved with the Japanese intelligence service had planned to blow up trains on the Trans-Siberian railway.)

The liberals began to get worried about the very serious deterioration that was affecting the whole country. At a congress held in Moscow the 300 delegates of the zemstvos and municipalities voted unanimously for a 'Manifesto of the Nation' which denounced 'the regime, the bureaucracy which is even a threat to the security of the throne'. They called for a freely elected assembly of representatives of the nation, abolition of laws restricting freedom, and entry into the government by men who believed in reforming the state. Nicholas II agreed to receive a delegation, headed by Prince Trubetskoy, consisting of fifteen members, 'in ceremonial dress and wearing white gloves'. He replied to them:

Dispel your doubts. My will – the will of your Tsar – to convoke the people's elected representatives is unshakable. They will be associated with the government's work. I follow the progress of this plan day by day. I believe firmly that Russia will emerge regenerated from the ordeal she is undergoing. Let there be, as in days of old, unity between the Tsar and all Russia, between me and the delegates of the land, unity which will be the foundation of an order built upon truly Russian principles.

But those undertakings did not appear in the official communiqué authorized by the censorship.

In his diary Nicholas wrote:

6 June (1905). Extraordinary heat, 23 degrees in the shade. After the report, I received at the farm 14 delegates from the zemstvos and towns nominated by the recent congress in Moscow. We lunched alone. During the day we enjoyed ourselves with the children at the seaside. They splashed and romped in the water. For the first time I bathed in a temperature of $14\frac{1}{4}$ degrees, cool but refreshing. We took tea under the sunshade. I received

Alexeyev and Taneyev [the head of the imperial chancellery]. After dinner
we sat on the balcony. Storms, accompanied by impressive lightning-
flashes, approached from the south-west.

Nicholas felt less affected by these political events than by the
military situation, being especially worried by the mutinies which
followed the one on the *Potemkin*. 'There have also been disorders
on the *Pruth*,' he wrote. 'Let us hope they have been able to keep
control of the other crews in the squadron! The leaders must be
punished harshly, and the rebels severely.' A few days later: 'May it
be God's will that this painful and shameful business will end very
soon.'

The defeat at Tsushima disheartened him more than anything
else: 'We have had very contradictory news about our fleet's encounter with the Japanese. Only our losses are mentioned, with nothing
said about theirs. Not knowing about that is terribly worrying.'
Three days later: 'Confirmation received of the terrible news about
the loss of almost all our ships in a two-day battle. Rozhestvensky
himself has been wounded and taken prisoner! It is a fine day today,
and that has made our sorrow the heavier.'

Since 18 February the country had feverishly awaited the conclusions of the Bulygin Commission, which was to determine the
way in which a consultative assembly, the imperial Duma, would be
convoked. The Tsar chose 6 August, the feast of the Transfiguration,
for the announcement. Of the demands which had been formulated,
only the principle of the secret ballot had been accepted and would
be guaranteed; but there would be neither universal suffrage, nor
equality of representation (voting by estates instead), nor direct
election.

The response was almost unanimous. 'It is a defiance by eunuchs.'
Most of the revolutionary parties – the Socialist Revolutionaries and
both factions, Bolshevik and Menshevik, of the Social Democrats –
refused to take part in this 'electoral comedy'. The Union of Unions
spoke of replying with a general strike and, after protracted discussion in the universities and the factories, such a strike began and
developed irresistibly. Nicholas notes: 'The strikes on the railways,

which began around Moscow, have now reached St Petersburg. Today the Baltic line stopped working. The persons to whom I was giving audience had difficulty in reaching Peterhof. To keep us in touch with the capital the Dozorny provides a shuttle service twice a day. A pretty time.'

The strike had suddenly hardened because it was believed that the railwaymen's delegates invited to St Petersburg to discuss problems connected with workers' retirement had all been arrested. That report was false. What was not a false report, though, was the order given by Governor-General Trepov to his soldiers: 'Fire without warning and don't spare the cartridges' (14 October 1905).

Three days later, faced with the rising tide of insurrection, and after the St Petersburg soviet's call 'to throw off the chains of centuries-old slavery', when the strikes were affecting the public services and even the Senate, and seeing that even Trepov advised concessions, and that he would not find another general capable of crushing the sedition, Nicholas at last listened to Witte and agreed to some of the points in his report. In this memorandum, the minister had pointed out

that the clashes with the police and the army, the bombs, the strikes, the events in Caucasia, the disorders in the schools, the agrarian revolts, and so on, are important not so much in themselves as in their effect on the rest of the population, which has not seriously opposed them. We must not close our eyes but must see that we have to ensure that these revolts do not surpass in horror anything known to history. Since February the population's aims are much more far-reaching than before.

Witte also urged Nicholas not to let the mere word 'constitution' plunge him into 'a sort of panic'.

Nicholas yielded, but on condition that his surrender be conducted with solemnity. He decided, therefore, to grant, by the manifesto of 17 October, the Duma and the freedoms so long awaited.

This Manifesto of Freedoms, as it has been called, made a threefold promise. First, to grant fundamental civil liberties to the population on the basis of these principles: effective inviolability of the person, freedom of conscience, of speech, of assembly and of association; second, without delay to facilitate participation in the Duma by the classes of the population at present without electoral rights, leaving it to the legislative authority to implement the principle of universal suffrage; third, to establish as an unchangeable rule that no law was to enter into force without the Duma's approval, and that the people's deputies would be given the opportunity to check on the legality of actions taken by the appointed authorities.

This manifesto embodied several of the demands put forward by liberal opinion and was hardly compatible with the principle of autocracy. Nicholas declared himself at once against universal suffrage: 'We must avoid taking too great strides. We should find ourselves, later, close to being a democratic republic. That would be senseless and criminal.' In the text of the manifesto, while nothing was said about autocracy, nothing was said, either, about a constitution, an amnesty or the ending of the state of siege. The chairman of the St Petersburg soviet, Leon Trotsky, could write:

We are granted freedom of assembly, but meetings are surrounded by troops. We are granted freedom of speech, but the censorship remains intact. We are granted freedom of education, but the universities are occupied by the police. We are granted inviolability of the person, but the prisons are still full ... We are granted everything and we have nothing ... We want neither the wolf's snout nor the fox's tail. The proletariat rejects the *nagaika* wrapped in the parchment of a constitution.

'The people have won, the Tsar has capitulated, the autocracy is no more,' observed *The Times*. For liberal opinion, indeed, the break with the former autocratic regime had been achieved, since an assembly was at last to meet, like the constituent assembly in France in 1789. These liberals who accepted the manifesto, and were thereafter known as 'Octobrists', failed to perceive that while winning one battle they had lost another, since the masses – or at least the organized workers – were breaking with the regime and preparing for a revolution which would not be merely a political change in the form of a more or less representative assembly.

Russia's history was thus at a turning-point. Up to this time, even if society was divided, the opposition to the autocracy had been united, under the guidance of the intelligentsia and the leaders of the nationalities. From now on the masses and the liberals diverged. Their aspirations had suddenly become different: social revolution for the former, a political adjustment for the latter.

Two men, at least, were quite well aware of this phenomenon. Milyukov, the leader of the liberal party of Constitutional Democrats, was one: he wanted the Duma to become sovereign. Lenin, the founder of the Bolshevik Party, considered that there was no longer any room for negotiation with the regime.

Like a film which stops and then starts again later, the Revolution of February 1917 began exactly at the point arrived at in October 1905: an assembly which was not sovereign, an insurrection in the capital with formation of a soviet, and leaders of the Duma trying to find a way of conciliation between the social revolution that they wanted to prevent and the political changes that Nicholas II was forced to accept. In 1917 he abdicated – but too late for the conciliators.

True, between 1905 and 1917 there had been twelve years wherein the autocracy had regained the upper hand and the Russian Army had been defeated a second time. So that, rather than being a 'dress rehearsal' for 1917, 1905 was the first stage in a process which suffered interruption. But, at the time, who could know that?

After his October Manifesto the Tsar broke with Pobedonostsev and (following, though not without grumbles, the advice of some of his family, especially his mother) entrusted to Witte the task of forming a government. This was a real innovation, for hitherto each minister had been responsible only to the Tsar. Witte was not of a repressive disposition, but he did need to have a strong man who would take the place of Trepov. He chose Durnovo, and this gave the liberals and constitutionalists a good pretext for refusing to collaborate with him.

The situation was still grave, and in December 1905 there were more than 400,000 workers on strike, barely fewer than in October. The Tsar intended to put a stop to these disorders, and ordered

Sologub, the Governor of Moscow, to 'answer terror with terror'. Terror was especially severe in the Baltic provinces, where the peasants had rebelled against the German barons who were their landlords. But the trial of strength had gone on for too long and, exhausted by the strikes, the soviets in St Petersburg and elsewhere closed down. The police arrested the 267 deputies of the capital's soviet at a meeting at its headquarters. At the same time the insurrection of Moscow was broken, after a week of barricade fighting; the 'credit' for that went to Admiral Dubasov. The number of victims was beyond counting.

In all this Witte stayed in the background, but he had lost any reputation he had with the liberals, as also with the Tsar: 'As for Witte, since the happenings in Moscow he has radically changed his views; now he wants to hang and shoot everybody. I have never seen such a chameleon of a man.'

In the country districts the movement had actually begun after the defeats in Manchuria and Bloody Sunday. A hundred landlords' properties had already been burned down in the Ukraine in 1902. During the 1905 Revolution, 979 suffered the same fate, and there were 846 other attacks on such domains. Then the movement declined as repression proved successful, in country as in town, after December 1905. Lenin commented:

The peasants have burnt about 2,000 residences of big landlords and divided among themselves the goods stolen from the people by the noble plunderers. It was, alas, a job not done thoroughly enough. They destroyed only one-fifteenth of the landlords' houses, only one-fifteenth of what they should have destroyed in order to sweep from the face of the earth the shameful thing called large-scale landownership.

The peasants demanded that the land be given to those who tilled it. Essentially aimed at the landowning nobility, the peasants' action also struck at the forces of order, but the clergy and the Tsar's person were spared. The muzhiks hated the nobles, the landlords, the bureaucracy, the 'gentry', but not the Tsar.

As for the army, made up of peasant soldiers, it fired when ordered to. Twelve years later, when the army rebelled in its turn and then refused to march against the Germans, Kerensky roughly reminded it: 'In 1905 you weren't afraid to fire on your brothers.' In 1917 they fired no more.

The situation in 1905 presents a really complex scene; for, while the army fired on the workers, it also mutinied. One hundred and ninety-three mutinies were recorded, 45 of them involving more than one unit, and 75 took place in Europe, far from the theatre of war. The mutinies occurred immediately after the issuing of the October Manifesto. The soldiers saw in it an invitation to voice their demands, relating mainly to excessive discipline and abuses to which this gave rise. However, their officers claimed that the manifesto had nothing to do with military order and discipline, or the citizens in uniform; it was not for them to take part in meetings or make speeches.

The mutinies stopped when the Minister of War, joining promises with repression, assured the soldiers that their pay would be increased and their food improved, that they would be given blankets and better boots, and so on. They stopped also because some officers who considered the lot of their men deplorable supported these demands and agreed with the minister's proposals. Thereafter the soldiers stepped back into line and obeyed the Tsar's orders, for 'without the Tsar the land is widowed and the people are orphans'. Only at Kronstadt did the amazing news circulate that a day would soon arrive when power would belong neither to the Tsar nor to God, but to a certain 'revolutionary committee' that remained nameless.

The army did not, however, experience such spectacular uprisings as took place in the navy. There were, to be sure, many units that joined in the revolutionary movement. In Moscow, for example, a 'military section' was even created in the soviet, with representatives from a dozen regiments. There were also large-scale military rebellions in Siberia and the Far East, but these were apparently connected more with the defeat and with the poor organization of the measures taken for the return of soldiers from the front than with any really revolutionary mood. At Irkutsk, however, in November 1905, a demonstration by 100,000 persons brought together railwaymen, workers, Cossacks and soldiers, a foreboding of February 1917.

The movement remained within limits because some of the officers in the army in the Far East were in sympathy with the soldiers, so that the latter did not suspect (as they were to suspect in 1917) that their officers were abandoning or betraying them. Twelve years on, what was at stake was something bigger, and bigger, too, was the anger of the common soldiers against their commanders. There was then no longer any solidarity between officers and men, except perhaps in the actual trenches. Mistrust had developed between the fighting men and the higher officers, and this led to mutinies that put an end to the repression ordered by the Tsar.

In 1906, nevertheless, the army command was fully conscious that if it wanted to keep control of the soldiers it would not do to use them against demonstrations. Rather than that, they should be trained to fight more effectively against the enemy. The point was made sharply by the Minister of War, General A. Rediger. When, at a meeting of the Council of Ministers early in 1908, he was asked how the training of young recruits was going, he pointed his finger at the Minister of the Interior, Stolypin, and said: 'The army isn't getting any training, because it's doing *your* job.'

And it did that job well, repressing and shooting. In 1908, though, the shame felt owing to the humiliating defeat by Japan was at its height. At the Council of Ministers, with the Tsar in the chair, General Rediger declared that the army was not ready even for a defensive war.

Nicholas II had assumed that the troubles would cease as soon as he convoked a Duma. Witte had 'promised' him that.

To that same Witte, though, he would have preferred 'an energetic soldier who would have crushed the rebellion by sheer force'. Trepov, at least, knew what to do. 'There's a man whom I can't do without,' said the Tsar. 'He is experienced, prudent, intelligent ... I give him a long memorandum, composed by Witte, to read, and he comes back soon to tell me in a few phrases what's in it.' There was also Orlov, of course, the general who had put down the revolt in Latvia, 'admirable work'. It was good, too, that the population had turned against the trouble-makers, who were all Jews. Nicholas wrote to his mother on 10 November 1905:

I hold a meeting of the Council of Ministers every week ... They talk a lot but do little. Everybody is afraid of taking courageous action. I keep on trying to force them – even Witte himself – to behave more energetically ... In your letter, my dear Mama, you ask me to show all my confidence to Witte; I assure you that I am doing my very best to ease his [*sic*] very difficult position ... But I must confess I am disappointed in him in a way. In everybody else's opinion he is a very energetic and even despotic man who straight away would try his utmost to re-establish order.

And on 12 January 1906 the Tsar wrote to the Dowager Empress about Witte: 'He is absolutely discredited with everybody, except perhaps the Jews abroad.'

For Nicholas, as for those around him, everything bad that happened was the Jews' doing. Even the difficulty experienced in raising loans in France was attributed to them. Wilhelm also put this idea to the Tsar: 'That the French refused a loan to Russia now has not so much to do with the Moroccan affair ... but [*sic*] to the reports of the Jews from Russia – who are the leaders of the revolt – to their kinsmen in France who have the whole press under their nefarious influence' (29 January 1906).

Whenever an assassination occurred, a Jewish revolutionary was accused. People would then cross themselves, saying: 'It couldn't have been a Russian.' Actually, the Socialist Revolutionaries who threw bombs were no more Jews than Plekhanov was, or Lenin. The man who killed Grand Duke Sergei gave his name as Bryusov, but none the less it was said that he was Jewish, and this was said of the perpetrators of other such deeds.

In the Ukraine as in Russia there had been pogroms before Alexander II's murder. Since 1881, however, the authorities and the police had encouraged them. For Alexander III and his ministers the Jews became a convenient scapegoat and anti-Semitism for the first time a means of government. Charges of ritual murder served as pretexts among the masses.

These pogroms were fostered through other accusations, especially religious ones. Children were taught by the priests that the Jews had killed Jesus, so committing the worst of crimes. There were also accusations connected with the economic activities of the Jews. As agents of the landed nobility (especially in the old Poland), as managers of mills, salt-works and taverns, or as money-lenders, they were made responsible for the peasants' poverty, their drunkenness and their turbulence. The peasants begrudged having to pay the Jews interest on the money they were obliged to borrow from them because the banks would not lend to the insolvent. Consequently, the authorities saw the Jews as the source of trouble – and even more so when Jews became active revolutionaries. The situation was so delicate that some of these Jewish revolutionaries did not dare to speak against pogroms, because they feared lest this might interfere with the first manifestations of political awareness on the part of the peasants, who would go on later, they thought, to turn against the landlords and the regime. For their part, the Tsarist bureaucracy considered that Jewish bankers were bleeding the Russian state by exploiting the Treasury in the same way that the 'small' Jews were impoverishing the peasants. The fact that a Jewish advocate, Crémieux, had defended Jews who were wrongly accused of ritual murder and that, at the same time, the Banque Crémieux hesitated to let the Russian state have a loan gave the Romanovs and the country's leading circles the idea of a 'Jewish plot'.

The Tsarist police even concocted, c. 1903, a forgery, the *Protocols of the Elders of Zion*, in which one could read that the founders of the Bund, the Jewish socialist party, had drawn up at a conference in Basel in 1897 a plan for world revolution. Ritual murders, secret intrigues, plots – one came across the Jews everywhere. In *The Brothers Karamazov*, Dostoevsky does not deny the truth of the 'blood libel'. Some Russians claimed that political assassinations were the up-to-date version of ritual murders.

Pobedonostsev, the theoretician of autocracy, had told the Tsar how he hoped to see the Jewish question solved in Russia: 'A third will emigrate, a third will convert to Christianity, and a third will die out.'

The revolutionary movement which appeared in 1899–1901 hastened the formation of reactionary and anti-semitic groups[1] like the Union of the Russian People, or the Black Hundreds, who took over the ideas Drumont had expounded in France, and promoted pogroms. In 1905 pogroms in Kishinev, Gomel and elsewhere continued a tradition inaugurated in Elizavetgrad in 1881. Everything happened under the eyes of the police, who were unwilling to intervene: 'They could not fire on Christians to protect Jews.'

Nicholas gave his views on this matter in a letter to his mother written in late October 1905:

I am thinking of having a heart-to-heart talk with you by making use of Isvolsky's return ... I'll begin by saying that the whole situation is better than it was a week ago ... In the first days after the manifesto the subversive elements raised their heads, but a strong reaction set in quickly, and a whole mass of loyal people suddenly made their power felt ... The impertinence of the Socialists and revolutionaries had angered the people once more; and, because nine-tenths of the trouble-makers are Jews, the people's whole anger turned against them. That's how the pogroms happened.

And the Tsar goes on:

It is amazing how they took place simultaneously in all the towns of Russia and Siberia. In England, of course, the press says that those disorders were organized by the police; they still go on repeating this worn-out fable. But not only Jews suffered: some of the Russian agitators, engineers, lawyers and suchlike bad people suffered as well. Cases as far apart as in Tomsk, Simferopol, Tver and Odessa show clearly what an infuriated mob can do; they surrounded the houses where revolutionaries had taken refuge, set fire to them and killed everybody trying to escape.

1. Anti-Semitism served as the basis for a pre-modern ideology, the Jews being, for the anti-Semites, the vanguard of a capitalist development that would lead to the country becoming democratized. Whereas Witte, who favoured economic progress, checked the adoption of anti-Semitic measures, the Minister of the Interior, representing the aristocratic order, promoted them (cf. H. D. Löwe).

I am receiving telegrams from everywhere with touching gratitude for the liberties conceded, but also many indicating that they want autocracy to be preserved. Why were they silent before, the good people?

'The good people' had been silent because for a long time and until now they had not known what to do. The formation of an extreme Right had been blocked by a difficulty of principle; to organize would have meant that the Tsar needed support, that the regime was weak. Moreover, as K. Golovin explained, 'the mere appearance of a political force, even an ultra-loyalist one, would have seemed a sort of mutiny'. Nevertheless, some traditionalists considered that it was no longer possible to rely on the bureaucracy; the liberal 'gangrene' had reached it, and the regime was in danger of rotting from within.

There was formed, towards the end of 1900, Russkoe Sobranie ('the Russian Assembly'), an intellectual club, which Plehve joined, as an ordinary member, after having at first wanted to suppress it. Presided over by Prince D. Golitsyn, it had about forty members: generals, high officials, jurists, publicists like A. Suvorin and V. Purishkevich. It spread to the provinces and its representatives affirmed to the Tsar in November 1904 their threefold faith: Orthodoxy, autocracy, fatherland. They opposed any shameful peace with Japan.

Asleep during the 1905 Revolution, though more or less continued by an Orthodox organization made up mostly of priests (Obshchestvo Khorugvenostev, 'the Banner-Bearers'), the Right was all at sea when faced by the revolutionary upsurge. On the initiative of a member of the State Council, B. Stürmer, it tried to work out a counter-programme to that of the liberals with whom the government was trying to compromise, but failed in the attempt.

The manifesto of October 1905 acted as detonator and resulted in the regrouping of an organized and aggressive extreme Right. The regime was showing weakness and it was no longer clear who was in charge. Was it Witte, the incumbent Prime Minister, whom the Right accused of treason for having made the Tsar issue the manifesto? Was it Durnovo, the Minister of the Interior, whom one would have wished firmer? Or General Trepov? It was necessary to organize, to find support among the masses, to counterbalance the revolutionaries. This idea, put forward by Dr A. Dubrovin, appealed to most members of the small rightist movements that had existed before the revolution. In this way the Union of the Russian People was born. It grew and developed like an avalanche. Figures vary, but within two years the Union had 3,500 centres of influence and at least 100 branches, with a membership estimated at between 600,000 and 3 million. With the help of the authorities it fomented pogroms against Jews and liberals. It had 2,200 members in Odessa in 1907. It had its own newspaper, *Russkoe Znamya* ('The Russian Standard'), and organized meetings and processions. It can be held responsible for the Beilis case, in which a Jew was wrongly accused of the ritual murder of a child.

On 23 December 1905 Nicholas II himself agreed to wear its badge. Later, defeated in the elections to the first Duma, the Union opposed Stolypin's policy and received the Tsar's official support in the 1907 elections. Far from placing himself above all the parties, Nicholas sent to the Union a telegram saying: 'May the Union of the Russian People be a support for me and an example to each and everyone of what is meant by law and order' (June 1907).

The Tsar counted on finding support in organizations like that. Meanwhile, however, all the élite of the zemstvos, the essence of educated society, wanted nothing better than to take part in the new regime that was due to emerge from the Tsar's October Manifesto. Its adherents, the 'Octobrists', represented the big industrial bourgeoisie that was developing, together with a section of the landlords. Their leader was A. Guchkov, a Moscow industrialist who played a central role in the Duma between 1906 and the fall of Tsardom.

'Our people have become free politically,' said the proclamation issued by the Octobrists,

and our state has become a law-governed state. A new principle has been introduced, namely, constitutional monarchy ... This new order opens up a new path for our country. The great danger which resulted from centuries of stagnation, and threatened our country's very survival, calls for unity

among us and for the formation of a strong authoritative government that
will be backed by the people's confidence – this alone can rescue our
country from the present chaos.

The proclamation continued with a whole programme of measures
intended to ensure civil rights, solve the land question, develop
education, improve the workers' conditions, and so on.

However, this enthusiasm on the part of the Octobrists was based
on a misunderstanding. Everything said in their proclamation was
felt by Nicholas as deeply wounding; he had certainly not wished to
establish a constitutional monarchy, but had merely yielded to the
pressure of events. To him, respect for a tradition did not mean
stagnation. And he considered illegitimate the right assumed by the
political parties to set forth programmes of legislation. It infringed
the principle of autocracy and his own sovereignty.

Actually, the provisions of the Octobrists' proclamation and the
terms of Witte's memorandum had something in common. The
minister had 'borrowed' some paragraphs of his memorandum
(which were to resurface in the October Manifesto) from a treatise
written by V. Kuzmin-Karavayev, a left-wing liberal.

Nevertheless, a section of the liberals rejected the October Manifesto. These bourgeois radicals had formed, a few days previously,
the Constitutional Democratic Party ('Cadets'), which held that
the government should be responsible to the Duma and enjoy its
confidence. They also demanded guarantees for civil liberty, abolition of the death penalty, abolition of the State Council, and,
above all, expropriation, with compensation, of part of the bigger
landholdings, so as to help the landless peasants.

Confronting these liberals, whom they called 'the bourgeois
parties', were the socialist parties – the Social Democrats and the
Socialist Revolutionaries. Ever since 1883, when G. Plekhanov
founded the 'Liberation of Labour' group, Russian Social Democrats
had always distinguished between the bourgeois–democratic revolution, which the fall of the autocracy would seal, and the socialist
revolution, which would come later, as the result of capitalist
development and the rise of a labour movement. In clearly separating

these two stages Plekhanov and the Marxists sought to mark themselves off from the populists who advocated the immediate establishment of a socialist order in Russia, through emancipation of the peasants and development of the village commune. However, given the rapid growth of the working class, the Social Democrats thought the interval between the two revolutions would not be a long one.

The events of 1904–5 made these analyses relevant in a new way, and the Social Democrats were soon obliged to discuss whether they would participate in a revolutionary provisional government, should one be set up. Martynov, Dan, Trotsky and Plekhanov rejected this line of action, unless the revolution were to spread to the rest of Europe, but their principle was challenged by Lenin, who considered that they made the mistake of mixing up the fight for a republic with the fight for socialism. The consequence was that, during the Revolution of 1905, Bolsheviks and Mensheviks differed from each other not so much over their organizational principles – a conspiratorial, vanguard party versus a democratic party – as over their tactics and objectives, with the Bolsheviks showing themselves both more radical and better disciplined.

In any case, after they had played a leading role in the soviets, Mensheviks and Bolsheviks split on the question of participation in the elections to the Duma. The Bolsheviks declared hostility to this 'masquerade', and had nothing but sarcasm for 'constitutional illusions that served only to prevent the workers from taking the road of insurrection'. The Mensheviks, on the contrary, took part in the election campaign in order to popularize their ideas. They won eighteen seats. The Bolsheviks denounced them but, a few months later, followed their example when the elections to the Second Duma were held, taking the line that this Duma could be used as a platform for their propaganda.

The Socialist Revolutionaries decided, like the Bolsheviks, to boycott the elections to the First Duma. However, faced with disappointment among the peasants, who were impatient to get a chance to put forward their demands, a section of the party broke away to form the 'Labour Group' (Trudovaya Gruppa), whose members called themselves Trudoviks, and included Alexander Kerensky.

Thus, the electorate were presented with an extreme Right, the Union of the Russian People, an Octobrist right-of-centre party, a Cadet left-of-centre party and a Left made up of Trudoviks and Mensheviks, while the extreme Left, consisting of SRs and Bolsheviks, boycotted the elections.

Given the electoral system adopted, which was based on a property qualification, and an unequal degree of participation, this election produced a crushing victory for the Cadets, who won 179 seats out of 486. The Octobrists had only 44, the extreme Right 100, the Trudoviks 94 and the Mensheviks 18. The parties representing non-Russian nationalities won about 100 seats.

In face of this relative repudiation – the government could count on no more than 144 votes, plus the votes of some of the 'nationalities' – Nicholas and Witte decided to strengthen the powers of the State Council, whereas the party with the largest number of seats called for its abolition. This was felt to be a provocation.

Strengthened by the successful repression in Moscow and the dispersal of the soviets, and encouraged, also, by the emergence of spontaneously organized forces favourable to autocracy, Nicholas assured one of these forces, the Union of the Russian People, that 'he intended to bear the burden of power alone ... The sun of truth would soon rise over the Russian land and dissipate all doubts.' He had, of course, been obliged to issue the October Manifesto, but he meant to restrict the powers of the Duma, which was soon to meet.

The Tsar was very alert and vigilant when the principles of his autocratic power seemed in danger of being challenged. His participation in and statements during the discussions on the Fundamental Laws, held at Tsarskoe Selo between 7 and 12 April 1906, testify to that. The central point being discussed was article 4 of the New Fundamental Law of the State, which was to replace the old article 1. The latter stated that 'The Tsar of All Russia is an autocratic monarch, with unlimited powers. He is to be obeyed not out of fear but as a matter of duty, in accordance with divine decree.' The new article stated: 'The Tsar of All Russia possesses supreme autocratic power. He is to be obeyed not out of fear but as a matter of duty, in accordance with divine decree.' The word 'supreme' (*ver-*

khovnaya) had been substituted for the word 'unlimited' (*neo-granichenny*). Here are the minutes of the discussion that took place:

His Imperial Majesty Nicholas II: Let us look at this Article 4. It includes an important change. I have thought a great deal about this matter since the Fundamental Law was subjected to review. It is a whole month since the Chairman of the Council of Ministers submitted the revised version to me. I am filled with doubt. Have I the right, before my ancestors, to alter the limits of the powers they bequeathed to me? This inner conflict still troubles me and I have yet to reach a decision. It would have been easier for me to make up my mind a month ago ... but since then I have received heaps of telegrams, letters and petitions from all parts of Russia and from persons belonging to all classes of society. They express their loyalty to me and, while thanking me for the October Manifesto, ask that I do not limit my powers. They want the Manifesto and preservation of the rights granted to my subjects, but are against any further step being taken which would limit my own powers. They desire that I remain autocrat of All Russia.

    I am, believe me, sincere when I tell you that if I were convinced that Russia wanted me to abdicate my autocratic powers, I would do that, for the country's good. But I am not convinced that this is so, and I do not believe that there is need to alter the nature of my supreme power ... It is dangerous to change the way that power is formulated. I know, too, that if no change is made, this may give rise to agitation, to attacks ... But where will these attacks come from? From so-called educated people, from the proletariat, from the Third Estate? Actually, I feel that eighty per cent of the people are with me. Nevertheless, I have to decide whether Article 4 is to remain as drafted.

I. Goremykin [whom Nicholas had just appointed in place of Witte]: Eighty per cent of the population would be worried, and many would be discontented if your powers were to be limited.

Count K. Pahlen: The question is whether we are to leave in the word 'unlimited'. I have no sympathy for the October Manifesto, but it exists ... Your Majesty cannot any longer make laws without reference to the legislative institutions. The word 'unlimited' cannot remain in the Fundamental Law.

M. Aksimov:... I am not an advocate of the liberties granted by the October Manifesto, but His Majesty did voluntarily limit his powers by issuing that document ... Putting back the word 'unlimited' would signify throwing down the gauntlet to the Duma ... It ought to be removed ...

Count D. Solsky: Perhaps the Fundamental Law ought not to be published.

Grand Duke Nicholas Nikolaevich: The word 'unlimited' was erased by Your Majesty in the manifesto of 17 October.

Grand Duke Vladimir Alexandrovich: I agree with my cousin.

P. Durnovo: The trouble comes not from the people but from the educated classes ... They it is who govern the state ... They are bound to consider that the term 'unlimited' contradicts the manifesto ... They will lead the people in revolt ...

His Imperial Majesty Nicholas II: I propose that we adjourn for fifteen minutes.

(*A quarter of an hour later ...*)

His Imperial Majesty Nicholas II: I will give you my decision later. Are there any other matters on the agenda?

(*Three days later, 9 April.*)

Count D. Solsky: Your Majesty expressed the wish to suspend your decision regarding Article 4. What is your desire – to retain or to exclude the word 'unlimited'?

His Imperial Majesty Nicholas II: I have decided to retain the wording of the Council of Ministers.

Count D. Solsky: Therefore the word 'unlimited' is to be deleted?

His Imperial Majesty Nicholas II: Yes – exclude it.

The New Fundamental Laws were published, accordingly, on 12 April, by the new Prime Minister, Goremykin. The Tsar, inviolable and sacred, was the sole initiator of laws. He was head of the armed forces, of the administration, of the diplomatic service. He alone conducted foreign policy and appointed and dismissed ministers, who were not responsible to the Duma, which was not empowered to intervene in military matters. The Tsar could dissolve the Duma but, by virtue of article 87, he could also, during parliamentary vacations, take 'any legislative measure demanded by exceptional circumstances'.

But his powers were no longer 'unlimited'.

The opening of the Duma on 26 April 1906 recalls that of the French States-General on 4 May 1789. Count Kokovtsov, Minister of Finance since 1904, has left an eyewitness account:

St George's Hall, the throne room, presented a queer spectacle at this moment, and I believe its walls had never before witnessed such a scene. The entire right side of the room was filled with uniformed people, members of the State Council and, further on, the Tsar's retinue. The left side was crowded with the members of the Duma, a small number of whom had appeared in full dress, while the overwhelming majority, occupying the first places near the throne, were dressed as if intentionally in workers' blouses and cotton shirts, and behind them was a crowd of peasants in the most varied costumes, some in national dress, and a multitude of representatives of the clergy. The first place among these representatives of the people was occupied by a man of tall stature, dressed in a worker's blouse and high, oiled boots, who examined the throne and those about it with a derisive and insolent air. It was the famous F. M. Onipko, who later won great renown by his bold statements in the first Duma and who also played a prominent role in the Kronstadt insurrection. While the Tsar read his speech addressed to the newly elected members of the Duma, I could not take my eyes off Onipko, so much contempt and hate did his insolent face show. I was not the only one who was thus impressed. Near me stood P. A. Stolypin, who turned to me and said: 'We both seem to be engrossed in the same spectacle. I even have a feeling that this man might throw a bomb.'

V. Gurko, an important personage in the higher administration, observed the same hostility, between the two parts of the chamber:

Naïvely believing that the people's representatives, many of whom were peasants, would be awed by the splendour of the imperial court, the ladies of the imperial family had worn nearly all their jewels ... This oriental method of impressing upon spectators a reverence for the bearers of supreme power was quite unsuited to the occasion. What it did achieve was to set in juxtaposition the boundless imperial luxury and the poverty of the people. The demagogues did not fail to comment upon this ominous contrast.

Nicholas advanced slowly and took his seat on the throne with a majestic air, while his servants arranged the skirts of his cape on the arm-rests. Beside him General Roop held upright the imperial sword. Opposite, Count Ignatiev held the imperial standard. On four pillars rested the crown, the sceptre, the orb and the seal. Silence fell, and the Tsar read out, very distinctly, the message a minister held before him. He had hardly finished when the orchestra began playing the imperial anthem, and a whole section of the assembly shouted 'Hurrah!' – which prevented I. Petrunkevich, F. Rodichev and others from replying to the sovereign's address. The ushers then escorted the participants from the chamber in perfect order and decorum. Outside, the sun was shining, and the regiments, in their uniforms of different colours that stood out against the blue sky, were lined up in impeccable formation.

At the very first meeting of the Duma conflict broke out at once between the assembly and the government. This conflict was to go on until 1917, regardless of the composition of the Duma, of whether or not the revolutionary Left had agreed to participate, of whether or not the electoral law excluded it, of whether or not the Duma met (if not, its reopening was demanded); for, however powerless it might be, the Duma was a platform, and the newspapers reproduced its debates.

At the opening session of this Duma, which was called 'the Duma of the people's hope', deputies Petrunkevich and Rodichev had one demand alone to put forward, namely: 'Free Russia insists on the release of all who have suffered for freedom.' It was this comprehensive amnesty that the Duma placed first in its address to the Tsar: 'We know how many crimes have been committed in the sacred name of the monarch, we know how much blood is hidden under the ermine mantle that covers the Emperor's shoulders.'

'Convict!' shouted a member of the extreme Right at Karaulov, who had, indeed, recently returned from Siberia.

'Yes, gentlemen, I have been a convict,' Karaulov replied. 'With my head shaven and chains on my feet I did tread the long road that leads to Siberia. My crime was this, that I wanted to make it possible for you to sit on these benches. I have made my contribution to the sea of tears and blood which has lifted you up to this place you now occupy.'

The keynote had been struck. A frenzied ovation saluted the speaker.

Embittered, the Tsar refused to receive a Duma delegation that wished to set before him a veritable programme of Western-style democratization – ministers responsible to the Duma, abolition of all the exceptional laws, guarantee of all freedoms, including freedom to strike, abolition of the death penalty, guaranteed employment for workers, free education for all, satisfaction of the nationalities' demands, drawing-up of an agrarian law to meet the peasants' need for land, and a full and absolute amnesty. After consulting Trepov and Stolypin, on 9 July 1906 Nicholas dissolved the Duma.

At that time the Grand Duchy of Finland enjoyed certain privileges, and the Tsar's police did not have the same powers there as in Russia. The left-wing deputies in the Duma – which meant mainly the Cadets, as most of the socialists had refused to participate in the elections – now went to Vyborg, in Finland, and from there issued a manifesto calling on the people to refuse to pay taxes or accept military service. When they returned to Russia they were arrested.

Stolypin, the new Prime Minister, was a strong man. Nicholas allowed him a free hand, commenting:

He offered ministries to Prince Lvov and Guchkov [one of the Octobrists' leaders]. I received them. They declined to serve, as also did I. Samarin, who had been proposed for the post of Procurator of the Holy Synod. Personal opinion predominates over patriotism with those people. It is associated with misplaced modesty and fear of compromising themselves. One must do without them.

To Stolypin he wrote: 'I have spoken with each man for an hour. They are not fit to be ministers. They are not men of action. That is particularly true of Lvov ... One must stop trying to make ministers of them.' And Stolypin stopped.

His daughter having been the victim of a bomb attempt on his life, Stolypin's repressive measures were all the harsher, and the new minister's reputation was soon established. The rope with which the rebels of 1905 and other terrorists were hanged came to be called 'Stolypin's necktie'. He governed by means of article 87 and specially created courts martial. But his grand idea was to queer the opposition's pitch by carrying through a large-scale agrarian reform.

At the elections to the Second Duma, held in February 1907, a marked slippage to the Left was observable. Having realized that the Duma provided a platform from which they could speak to the people, Lenin's Bolsheviks took part in the elections, along with the Mensheviks and SRs. To the left of the 94 Cadets there were now 118 socialists, plus a substantial group of Trudoviks. Using the excuse of a Social Democratic plot against the Tsar's life (whereas, actually, it was the SRs who were involved), and the Duma's refusal to waive the parliamentary immunity enjoyed by its members, Stolypin at once dissolved the Second Duma, which would clearly have been uncontrollable, and he had the Social Democrat deputies arrested.

A new electoral law was prepared, violating the Fundamental Laws that had been proclaimed in 1906. This 'coup d'état' made possible the appearance of a Third Duma, which came to be nicknamed 'the Duma of the lords, priests and lackeys'. It was to remain in session till 1912. 'Thanks be to God', a minister was able to say, 'in Russia we have no parliament!'

Since he did not want an assembly made up of 'professors', but one of 'sound men, well rooted in their respective provinces', Stolypin contrived to restrict entry to this Third Duma to the direct representatives of only seven cities, instead of the twenty-five that had been represented in the previous legislature. Of the deputies as a whole, 44 per cent were nobles. Only 3,500,000 people had been allowed to vote. The number of deputies from the non-Russian nationalities was cut down: 14 Poles instead of 35, 10 from Caucasia instead of 25, and so on. All the same, given the repression that now got under way, and with the number of citizens who were *not* living under the jurisdiction of some form of 'exceptional measures' limited to 5 million, this Third Duma was 'a new oasis of political and academic freedom' (T. Riha) in Russia's political desert.

However, this 'oasis' had few means at its disposal. The Tsar still retained an absolute veto over all legislation; he decided foreign policy; and it was open to him to use article 87 without consulting the Duma, thereby recovering his powers in their entirety. The Duma could talk, of course, but its very composition ensured that it was quite docile. Only a few speakers, such as the Octobrist Guchkov, criticized the way national defence was organized.

Even so, Nicholas could not tolerate such an assembly. Guchkov's attacks were indirectly aimed at the imperial family, since most of the higher commands were held by Grand Dukes. The Tsar refused to allow his ministers to discuss their plans before this assembly, and dismissed two of them, A. Rediger and A. Polivanov, for being too 'co-operative' with the Duma and entering into arguments about problems of defence.

In 1909 Nicholas forbade his new Minister of War, Vladimir Sukhomlinov, even to enter the Duma. 'Why do you discuss with them? . . . You are my minister, not theirs. I did not create the Duma in order to receive instructions but only to hear views.' Accordingly, Sukhomlinov refrained from going back to the Duma.

The Tsar was not alone in this hostility to all notions of representation. His ministers and officials also considered that all the good elements in the Russian nation were already employed in the state apparatus. Outside of state service were 'only a few Utopian or fanatical politicians'.

Now, the defeats suffered in the war with Japan meant a tragic setback for these 'best' men. The question needed to be re-examined. A certain attitude had long become engrained, however, and the only schools of government that existed at that time, the zemstvos and municipalities, were under surveillance. Between 1900 and 1914 the Minister of the Interior annulled the election of 217 mayors and municipal councils, out of the 318 towns that possessed these institutions. There was no question of discussing public affairs with 'those people'.

A kind of social gulf was thus created between the aristocracy and the bureaucracy, on the one hand, and the people, on the other, which was not bridged by the merchants or the industrialists, for they had grown used to bowing low before the authorities and were not interested in politics. It was this gap that opened the way for the 'working class', or rather for the political parties that claimed to speak in its name.

The parties, too, were under 'intensified' surveillance. Even after 1906 the only political groups enjoying legal status were the Union of the Russian People, the Octobrists and the party of 'Peaceful Reconstruction'. The socialist mass parties, whether 'worker', 'peasant' or 'bourgeois', lacked even the right to register themselves as such. Had not the Minister of the Interior, P. Durnovo, said: 'For my part, I don't recognize any political parties'? Between 17 October 1905, the day of the manifesto, and January 1906, 45,000 persons were sent into administrative exile for political reasons.

The only public place where a certain freedom of speech could always be exercised was, paradoxically, the court room, and this explains the special role played by advocates and jurists in the revolutionary movement – men like V. Maklakov, Kerensky and Lenin. Nevertheless, the press was freer after 1906 than before. The number of newspapers increased from 123 before the revolution to 800 in 1908 and 1,158 in 1913. Above all, the debates in the Duma, which were reproduced in all these papers, provided extraordinary resonance for the discontent that existed throughout the empire. Lenin appreciated this and persuaded the Bolsheviks to put up candidates at the elections to the Second Duma and make use of this platform. As only 6 per cent of the deputies to the First Duma were not unelectable as a result of the changed electoral law, this Duma, called 'the Duma of the people's anger', certainly allowed the revolutionary organizations to express themselves; but that situation did not last long, as Stolypin decided to dissolve it out of hand.

In the Third Duma they talked about everything – education,

reform of the Orthodox Church, the problem of the non-Russian nationalities (especially), the organization of national defence ... The Octobrist Guchkov continually rejoiced in the fact that this was now possible, thanks to the October Manifesto, which had turned Russia into a sort of constitutional monarchy. That was not exactly true, but it was true enough to irritate Nicholas greatly. The same applied to the Cadet Milyukov, a professor of history, who throughout ten years expressed the wish that the government be one that 'enjoyed the confidence' of the Duma or, still better, was responsible to that assembly.

'But when are they going to shut up?' demanded Nicholas.

All this concerned only the divorce between Tsardom and its 'legal' opposition. With the 'illegal' opposition – the Social Democrats, both Mensheviks and Bolsheviks, the Socialist Revolutionaries, Kerensky's Trudoviks, and so on – there was no question of dialogue. All their leaders had had to emigrate – Lenin was in Switzerland, Trotsky in France – unless they were in Siberia, like Breshko-Breshkovskaya, Tsereteli, Chernov and others. The trial of strength would be renewed, but with other actors.

Were people wrong about the character of Nicholas II? He was considered to be indecisive, a playboy, gentle, open to influence – in short, a bit simple. And the enemies of the regime were always blaming his ministers or the people round him. In 1906–7 the Tsar was even credited with an attempt at an 'approach' to the Duma deputies. But the Cadets, headed by Milyukov, had been uncompromising, demanding that their programme be accepted in its entirety.

'Wait, be more patient,' they were advised by Mackenzie Wallace, an Englishman who had known Russia over a long period.

'Wait for how long?' asked one of the Cadet leaders, in reply.

'Eight to ten years. In England it took us a century to arrive at a parliamentary monarchy.'

'Eight to ten years is much too long. We shall never wait so long as that.'

Nicholas II had sensed perfectly well the uncompromising attitude of those deputies, and took note of it. 'Far from finding him distressed, dejected, suffering for his people,' Prince Lvov tells us, 'I stood before a cheerful fellow dressed in a strawberry-coloured jacket with a broad band across it' – the uniform that he had just presented to the imperial family's own battalion of riflemen.

To V. Kokovtsov, Nicholas said:

I never intended to embark upon that distant and unknown journey which I was so strongly advised to undertake ... I wished to verify my own thoughts by asking the advice of those whom I trust ... I have no more misgivings nor have I ever really had them, for I have no right to renounce that which was bequeathed to me by my forefathers and which I must hand down unimpaired to my son.

And, according to Kokovtsov, he had told Stolypin,

that in his estimation the dissolution of the Duma had become a matter of urgent necessity and could not be postponed any longer. 'Otherwise,' he said, 'we all, and I in the first place, will bear the consequences of our weakness and hesitation.' He went on to say that no one could tell what would happen to the country unless something were done to remove this source of direct instigation to mutiny ... 'I am obligated before God, before my country and before myself to fight, and I would rather perish than hand over without any resistance all power to those who stretch out their hands for it.'

But, in reality, the circle was closing in around him. Faced with the intransigence of the regime and its refusal to engage in dialogue with the country, terrorism had raised its head again. At first, those who had practised terrorism had thought that the lesson of the revolution would be enough and Tsardom would reform itself.

Beheaded under Alexander III, the terrorist organization had by now been reorganized, but cautiously, for those concerned knew that their ranks had been infiltrated by the Okhrana. Assassinations had nevertheless been resumed. In 1901 the student P. Karpovich struck down the Minister of Education. It was the act of a loner, but carried out in accordance with moral principles: 'My death will serve as expiation for the crime I have committed.' *The Times* wrote, on the morrow of Plehve's assassination in 1904: 'He carried autocratic theory and practice to lengths which even in Russia seemed abnormal. He shut every tap and screwed down every safety-valve; and at last the world learns without surprise that the boiler has burst.'

Terrorism at that time found its justification in a regime that was regarded as iniquitous, and democratic and liberal opinion abroad approved of it. 'The bomb is the only means by which rebellious opinion can make itself heard.' This justification was accepted because, since Russia had no revolutionary bourgeoisie, it was necessary for workers, peasants and intellectuals 'to force out the rusty nails that keep our coffin shut' (Gershuni). Terrorism struck many times: Bogolepov, Sipyagin, Plehve, then Bobrikov, Grand Duke Sergei, General Kozlov (killed by mistake for Trepov), General Min and, finally, Stolypin.

Since the Revolution of 1905 and the cruel repression that followed it (not to mention the unnumbered victims of Bloody Sunday), the Tsar was no longer called Nicholas 'the Unlucky' but, instead, Nicholas 'the Bloody', and the SRs' terrorist organization thenceforth took him as its target. The Okhrana knew about that, for the head of the organization was none other than Azef, a double agent who acted as informer for Gerasimov, the chief of police, while at the same time performing great services for the revolutionaries. On two occasions attempts on the life of Nicholas II failed. A sailor who was supposed to kill him admitted that, at the last moment, his hand started to tremble.

'This rabble must be liquidated,' was Nicholas's refrain. He refused to intercede on behalf of the revolutionaries. 'And they dare to talk to me about an amnesty!' he said one day, banging his fist on the table.

Terrorism had its victims. What was the record of repression? Here is the report prepared by Karl Liebknecht for the congress of the Second International planned for August 1914:

Statistics that are as yet incomplete, based on official sources, give the following picture:

The number of persons sentenced to death for political 'crimes' between 1906 and 1910 was 5,735, which accounted for nearly one-sixth of all who were brought to trial on political grounds. Of these 3,741 were executed.

The atrocious character of these figures stands out especially when we consider that during the period 1825–1905, that is, the 80 years preceding the revolution, only 625 'politicals' were sentenced in Russia and only 192 executed.

In the first five years of the constitutional era the number of death sentences was 180 times that number, and the number of executions 250 times. In recent years the number of executions in Germany averaged about 15 per year.

Between 1906 and 1910 the courts condemned for political offences 37,620 persons altogether, 8,640 being sent to *katorga* (forced labour), 4,144 to prison, 1,292 to disciplinary battalions and 1,858 to compulsory settlement. Every person so condemned was also deprived of all civil rights.

'Compulsory settlement' means deportation of persons into inhospitable deserts, without any aid. It is a method that differs little from that employed by the Young Turk regime to get rid of the stray dogs in old Constantinople. The 'settlement' areas are among the least fertile and most icy in the world. A temperature of 30 to 50 degrees below zero prevails in many of these places during several months of the year. There the 'settlers', reduced by force to a condition of savagery, have to struggle for their wretched existence with the most primitive means and without a kopeck of assistance. Among them there are women and children. Often this penalty is inflicted for mere membership of the Social Democratic Party. Today there are between 5,000 and 6,000 of these 'compulsory settlers'.

To the condemnations by the courts must be added an immense number of sentences to prison or exile imposed by the administrative authorities. The prisons and places of custody, the most grimly famous of which are those at Serentui, Akatui, Tobolsk, Orel, Yaroslavl and Moscow (Butyrki), provide today, according to official estimates (which ignore the minimum requirements of prison hygiene), 'places' for about 140,000 prisoners, that is, for nearly 50 per cent more than was the case three or four years ago. In 1913 there were, on average, about 220,000 persons, sometimes as many as 250,000, incarcerated there, and since then the number has increased still further, despite the famous Jubilee amnesty (on the occasion of the tercentenary of the Romanov dynasty), which benefited only the ordinary, non-political offenders. The prisoners are often packed in more closely even than cattle in their sheds, so closely, in fact, that sometimes they have to take turns to lie down to sleep. For much of their time in detention *katorga* prisoners are fettered, day and night, in iron chains. Quite commonly the piece of leather placed under the chains is removed so that the metal, rubbing against bare flesh, flays the skin.

For food the prisoners are allowed, on average, ten kopecks per head per day. This is, of course, far from sufficient to nourish a man, especially one who is living in such abnormal conditions, external and internal, as these Russian prisoners. Much even of this derisory sum remains, moreover, in the hands of that gang of robbers, the Russian bureaucracy. For what is left the prisoners can all too often obtain only wretched food prepared in a way that defies description.

Clothing, reduced to rags and filthy, is inadequate from all points of view. The most elementary requirements of cleanliness and hygiene are neglected. It seems incredible that human beings can live, even if only for a few weeks, in such deprivation and in such an atmosphere (it is often forbidden to open the air-vents). The right to exercise is systematically encroached upon or even abolished altogether. In most cases the prisoners are not given work – without which any loss of liberty becomes, even if living conditions are favourable, a punishment beyond endurance. Work is available only in the most painful and unhealthy forms, such as oakum-picking. In those cases where political prisoners had the right to an occupation of their choice, this has mostly been abolished.

In view of all this, the health of the prisoners cannot but be frightfully bad. Phthisis and dysentery, typhus and scurvy inflict deadly ravages. The death rate exceeds all limits. 55 per cent of deaths are due to phthisis.

But the barbaric treatment does not end there. Prisoners, particularly the politicals, are methodically degraded by being put in the same cells with common criminals and are often left to the dictatorial violence of the basest of these, the 'Ivans' who are evidently the favourites of a prison administration of this sort. Gross insults and humiliations are the politicals' daily fare: ill treatment accompanies them from morning till night, from the very start of their imprisonment. Hanging over their heads all the time is the threat of a barbarous disciplinary system in which the dark cell and beating, regarded by the Ministers of Justice and Police as being once more indispensable, are the principal methods used. Medieval-type tortures are on the menu in many prisons. This is how they strangle whatever is left of human sensitivity and dignity in those prisoners who do not succumb to epidemics or to the bullets of the warders stationed outside the windows of the cells and always ready to shoot. The only escape for the unfortunates who seek to leave this hellish existence is into death; and so, besides epidemics of diseases, there are veritable epidemics of suicide.

The Revolution of 1905 had been felt by the intelligentsia as a tragic defeat, and several years were to pass before it was seen as the first stage in a process, or even as a 'dress rehearsal'. Nevertheless, Merezhkovsky warned the West that, 'whether or not it has been successful, the Russian revolution is henceforth just as absolute as the autocracy it repudiates.' He took up Herzen's idea that Russia would never know a 'happy medium', and that it would 'enkindle a Europe that tried to extinguish it'. The intelligentsia's exasperation found expression in every form. Its target was not only the personnel of the regime and its State Council – painted by Repin like a still life, 'dead as Carthage, before being destroyed' – but, above all, Nicholas himself, depicted in caricatures as an ass. Poets like Konstantin Balmont voiced most forcibly the people's anger:

> Our Tsar
> Our Tsar means Mukden; our Tsar is Tsushima.
> Our Tsar is a bloody stain,
> A stench of gunpowder and smoke.
> Black is his soul.
> Our Tsar, sickly and blind,
> Is prison and knout, shooting and hanging,
> Tsar! You are the gallows-bird ...
> The hour of retribution awaits you,
> Tsar, who began where? At Khodynka.
> And will end where? On the scaffold.
>
> (1906)

Whereas until this time the intelligentsia had thought it unworthy to take to the streets and 'carry the red flag', they consorted now with terrorists: Merezhkovsky with Savinkov, Bely with Valentinov, Gorky with Lenin. Ever since the socialist parties were formed there had been taking place a sort of fusion between the strictly revolutionary section of the intelligentsia, on the one hand, and the writers and theatre people, on the other.

The freedoms acquired in 1905, especially freedom of the press, made it possible for opposition publications to expand their scope, and this was one of the chief differences between the situation before and after 1905. Before 1905 the fate of Russia and of Tsardom was discussed between individual thinkers who spoke in their own respective names and alone embodied opinions and strategies. After 1905 these same discussions, now often more radical in content, were carried on by an entire milieu, social and professional, of writers, artists and political activists, all mixed up together.

A whole 'literary civilization', to use Merezhkovsky's expression, was henceforth engaged in open political struggle against Nicholas II, joining in the fight alongside the organizations and parties. Men as different as the author of *The Golden Cockerel*, Rimsky-Korsakov, the painter Repin, Alexander Blok, soon to be the poet of the revolution, and the novelist Maxim Gorky thus found themselves together against the autocracy. However, the arrival of Gorky marked a shift and a fracture in this general movement of condemnation of Tsarism. For the Bolsheviks, towards whom he tended, he embodied the people, and for Lenin his reactions revealed the moods of the oppressed classes. But Gorky was linked not so much with the peasants as with the new urban strata to whose anger, impatience and need to escape he gave voice. 'Let the storm break!' demanded this singer of the tramps. The mysticism that inspired Gorky associated him with those who were worried by the mechanistic scientism of the new nihilists, the Marxists, and he participated in the religious and critical renaissance heralded by Rozanov, which led to the movement incarnated in *Signposts* (*Vekhi*). The terrorist assassinations were as repugnant to them as the crimes of Tsardom, and they considered that the end does not justify the means.

'Why should we get angry about the burnings-alive ordered by the Inquisition?' Rozanov asked. 'Those men, too, wanted to prepare a happy future for the generations to come.'

Down to 1905 Nicholas II and the court accused 'society' of spreading rebellion and stirring up the people against their beloved Tsar. As that myth collapsed another illusion began to appear – the intelligentsia's confidence that it was closer to the people than to Tsardom. Writing under his pseudonym 'Varvarin' in *Russkoye Slovo*, Rozanov described with ferocity the embarrassment experienced by Russia's liberal intellectuals: 'Having thoroughly enjoyed the gorgeous spectacle of the revolution, our intelligentsia prepared to don their fur-lined overcoats and return to their comfortable houses, but the overcoats were stolen and the houses burned.'

This nightmare, premature in 1905, but so real a few years later, was visualized by only a few, and quickly faded.

Actually, the repression, far from crushing the liberals and revolutionaries, had merely intensified their fury. They were no longer just a vanguard, as they had been before 1905, but had been doubly reinforced – the former as a result of the extraordinary economic development in progress, the latter in consequence of the negative effects of Stolypin's laws. His agrarian law, although it created the beginnings of an independent peasantry, brought about also an influx of the poorest peasants into the towns, where they formed a proletariat that was soon exasperated by its conditions of life and made capable of great solidarity through its concentration in giant factories.

The massacre of the workers in the Lena goldfields in 1912 gave the impetus to an extraordinary revival of the strike movement throughout the country. The strikes were concerned with wages, of course, but they increasingly assumed the form of revolts against the constraints and methods of control imposed on the workers in their workplaces. The Tsarist system, which hitherto had been felt as directly oppressive only by the élites (and by the oppressed nationalities), was now resented by these millions of Russians who discovered in the factory what it really meant, being enlightened by the propaganda and agitation of the revolutionaries, which were aimed at the administration, the government, the state.

Nicholas, angered by this political climate and increasingly under Alexandra's influence, tended to bring together in his mind phenomena that were actually different. When, for example, he learned that in London the King and the House of Commons had held a reception in honour of the Speaker of the Duma, he wondered if he ought not, by way of reprisal, to receive a delegation from Sinn Fein. In his mind, the opposition began with the Prime Minister. After Witte, men so little to be suspected of liberal ideas as Trepov and Stolypin were victims of his suspicion. 'God, what a time they took to dissolve the Duma! When will they all be made to shut up?'

For, thanks to the Duma, politics had become the common property of all citizens. But Nicholas would not hear of that. Even the extreme Right felt affronted by the Tsar's refusal to honour it, by taking a minister only rarely from its ranks. Although the Duma contained between 400 and 500 deputies who thought of themselves as the country's élite, or at least its representatives, where did the Tsar go to find a minister? Goremykin said of himself: 'I am like an old fur coat. For many months I have been packed away in camphor. I am being taken out now merely for the occasion; when it is passed I shall be packed away again till I am wanted the next time.' Called to office again in 1914, Goremykin was seventy-five years old. But he had the bearing of a higher civil servant who respectfully receives orders and passes them on. There was no danger of him offending his sovereign. And, besides, he too took no notice of the Duma.

The Fourth Duma, elected in 1912 for a term of five years, was still more conservative than its predecessors. Instead of the 190 peasants who had sat in the first Duma there was an absolute majority of nobles, together with 48 clergy. Right, Centre and Left balanced each other – but only if it be accepted that 'the Left' meant, mainly, the Cadets. The extreme Left, whose power was demonstrated in the strikes of 1905 and then again in 1912–13, had only 15 deputies. The more radical the country became the more the electoral law moved the Duma to the right and the greater grew the influence of the nobility.

Furthermore, the Tsar and the government refused even to exchange views with the Duma. A motion signed by the assembly shows the tone that prevailed in relations between it and the government. The motion was voted when the Minister of the Interior, N. Maklakov, himself a member of the Union of the Russian People, presented his budget – and this although the assembly broadly agreed with his ideas.

Whereas (1) the Minister, by continuing to employ exceptional measures, although order has been restored in the country, is causing widespread discontent among the population and a justified feeling of revolt against measures seen as unnecessary; (2) although the Government's authority should be based on the application of the laws, the Minister is mocking the law and destroying the population's respect for it; (3) by postponing the reform of local self-government the Government is hindering economic and cultural progress; and (4) by maintaining the present laws affecting the nationalities the administration is breaking up national unity and weakening Russia ...

'Isn't this inadmissible?' Maklakov asked the Tsar.

'Let the majority and the minority submit their views and the Tsar will decide.'

The Tsar approved this suggestion and, in June 1914, according to the Minister of Justice, Shcheglovitov, a special meeting of the Cabinet discussed the transformation of the Duma from a legislative body into a consultative assembly. Imagining the explosion that might follow, the Cabinet opposed the idea, but Nicholas's intentions had none the less been expressed, and clearly.

On the political plane, however, Stolypin had tried to create a sort of link, or at least a relay station, between the government, society and the representative institutions. He sent the Tsar a report on how the administration might be renovated (May 1911). Three measures were proposed: a Ministry of Local Administration to be created, to deal with zemstvo problems; the existing ministries to be strengthened by the addition of new departments concerned with health, labour, the nationalities and natural resources; and these ministries to be assigned to specialists, which would have meant restricting – *de facto*, but without saying so – the sovereign's ability to choose the members of his government. Stolypin is said to have proposed, even, that the Third Duma be dissolved and article 87 be brought into effect, which would signify that he meant to ride roughshod over the country's elected representatives. This was alleged, at any rate, by A. Zenkovsky, who was responsible for working out the details of Stolypin's plan. But nothing came of it, because Nicholas did not reply.

Stolypin's cleverness thus remained sterile. To be sure, he had succeeded in maintaining the alliance between the government and the Octobrist Right, which correctly saw in the State Council (which it had at first denounced) a possible counterweight to the Duma. By paralysing the action of the liberals, however, especially through his agrarian reform, Stolypin's policy had the effect of pushing the opposition leftwards, bringing some of the Octobrists closer to the Cadets, and the more radical of the Cadets closer to the more moderate socialists, while the left wing of the socialists became in its turn more radical than ever.

The general movement of *levenie* ('shifting to the left') echoed the growing discontent in the countryside, among peasants resisting the transformation of their village communes and also peasants who had been obliged by these transformations to leave their villages. The countryside was indeed relatively quiet after 1907 and quieter still after 1911, but social differences were increasing there, and in 1917 the negative effects of Stolypin's reforms found violent expression.

The Tsar's political mood was again made apparent during a conflict over the budget for the navy in 1909 and again during a constitutional crisis in 1911. On the first of these occasions the Duma had, after prolonged negotiation, submitted a bill to the government that dealt with the appointment and promotion of the members of the General Staff; its target was the way in which the Tsar gave commands to members of his family.

Nicholas wrote thus to Stolypin on 25 April 1909:

Pyotr Arkadievich:

Since my last conversation with you I have thought continually about the lists for the Naval General Staff.

Now having weighed everything, I have decided not to confirm the legislative project submitted to me ...

This is not a question of confidence or lack of it: it is my will.

Remember that we live in Russia, not abroad or in Finland (Senate) and, therefore, *I shall not consider the possibility of any resignation.* To be sure, there will be talk of this at St Petersburg and at Moscow, but hysterical shouting will soon subside ...

I warn you that I will reject categorically your petition or that of anyone else for resignation.

Sincerely, Nicholas

In the constitutional crisis of spring 1911 the Tsar was again opposed to Stolypin. Once more the question had arisen of introducing zemstvos into the nine western provinces, which Stolypin knew well through having held appointments there. The Right in the State Council objected on the grounds that one effect of the proposal would be to bring in too many Polish landlords. A fresh demarcation made it possible to cut down the number of these Poles (who were, however, supporters of the regime), and at the same time representatives of the clergy were added to the eligible groups, and the Jews excluded. Since the Tsar had already vetoed a reform of Stolypin's which was less favourable to the Polish nobles, the government retreated in order not to surrender, and the bill was welcomed in the Duma, being presented as favourable to officials who were Russian, and to Byelorussian peasants, and unfavourable to the Polish landlords and Jewish merchants. The State Council objected, saying that if the loyal Poles were excluded they would turn into opponents. In their argument both sides were pretending, because, in fact, the State Council was not favourable to the Poles as such but to the nobility, and the government was concerned to favour the peasants, regardless of whether or not they were

Byelorussian. Moreover, the State Council was dominated by Durnovo and Trepov, Stolypin's rivals. They secured the rejection of the bill. Stolypin told the Tsar that unless article 87 of the Fundamental Laws was brought into force, to overcome the council's opposition, he would resign.

Nicholas was circumspect. He sympathized with Trepov and Durnovo. Stolypin offended him by his authoritarian ways and also by the general respect he enjoyed. In the Duma they were starting to protest against improper use of article 87 – that is to say, the deputies were going to oppose something they had themselves voted for, in order to defend the rights of the assemblies. Would the Tsar sacrifice Stolypin to them? Would he make this concession to the parliamentary system?

Grand Duke Nicholas and the Tsar's young brother, Grand Duke Michael, favoured keeping Stolypin. The Dowager Empress was of the same mind:

I am fully confident that the Emperor cannot part with Stolypin because the Emperor himself must realize that part of the guilt for what happened belongs to himself ... If Stolypin insists on his own way, I do not doubt for a moment that, after long hesitation, the Emperor will end by giving in ... The more time that passes, the greater and deeper will grow the dissatisfaction of the Emperor with Stolypin. I am almost certain that poor Stolypin will win this time, but it will not be for long, and we shall soon see him out of office.

The Tsar gave in to Stolypin. He did not have to dismiss him: a few months later, the minister was murdered.

Stolypin's regime was not restricted to a policy of repression, the most severe that Russia had known. He was simultaneously promoting industrial development, following where Witte had led, and economic growth had never leaped forward as it did between 1906 and 1913, by 30 per cent and 150 per cent, depending on the sectors.

Undoubtedly, Russia was making her economic take-off, but she was still backward. The value of industrial production in 1913 was two and a half times less than in France and six times less than in Germany. Nevertheless, Russian capitalism was being born, and this had the effect of associating new social groups with the regime. Whereas Witte had mainly confined himself to economic considerations, Stolypin was above all a politician, and he therefore directed his attention to the peasantry. He aimed to improve their lot and thereby to defuse their revolutionary potential. In 1905, despite the agrarian revolts, the peasants had not shown any particular enmity towards Tsarism. Stolypin sought to play upon their loyalism and their land hunger, hoping in this way to win their support for the regime.

At this time discussion about the agrarian question was in full swing. Lenin and the Bolsheviks called for nationalizing the land, a programme that was fairly close to that of the moderate-socialist Trudoviks. The Mensheviks were for municipalizing the land. The Socialist Revolutionaries advocated dividing up the land equally, taking account of the number of persons in each household. The question that separated them most was whether or not the expropriated landlords should be compensated. Stolypin's idea, which was inspired by the writings of C. Koefoed, a Dane who enjoyed the advantage of the special connection between the courts of Copenhagen and St Petersburg, aimed at developing individualism and capitalism among the peasants by breaking up the village commune. The law would henceforth authorize each head of a household to require the transformation into private property of the share of land that the commune had assigned to him. The commune would not be allowed to deprive of his grazing rights the peasant making this request. If two-thirds of the members of the commune put in for this change, all the village land would be turned into private property. Naturally, most of the peasants, the less well off, opposed the change. Many of them sold their share and went off to settle in Siberia. What the peasants were more interested in was the uncultivated land in the possession of the big landlords, the state and the Tsar. Their discontent was expressed in their votes at the elections to the Second Duma and its successors.

Stolypin's reforms had nevertheless divided the peasantry by helping a large number of muzhiks, the future 'kulaks', to free themselves from the commune and by promoting the departure from their villages of another minority. Politically this was a success that worried the revolutionary leaders, Lenin especially.

In the Second Duma, called by the Right 'the Duma of the people's ignorance', because it included so many opponents of the regime, discussion went on and on about the problems involved in expropriation (while they were in progress the police arranged for two Jewish advocates of the measure to be murdered), and Nicholas could not stand theoretical arguments. Was he really interested in the reforms? Koefoed, who was presented to the Tsar by A. Krivoshein, the Minister of Agriculture, tells us:

He seemed to be more interested in my family history. He knew the Danish name Koefoed from his visit to Denmark and he evidently knew the late Admiral Koefoed personally. At any rate, he only questioned me on my kinship, for example, with the admiral. I had to disappoint him by saying that we must belong to the same family, but to different branches.

And that was what the Tsar's interest in Stolypin's reform amounted to.

Bloody Sunday had rent the sacred bond between the Tsar and his people, but only in the towns. It produced no echo in the rural areas. Nor did the few punitive expeditions that were sent into those areas later on have any effect; they were part of a tradition.

On the other hand, the dissolution of the First and Second Dumas seriously affected the country people, because they had placed all their hopes on the motions addressed to the Tsar, and now Nicholas was dissolving the assemblies that gave expression to their demands. The disaffection thus caused was intensified by Stolypin's reform. In his attempt to rally the dynamic element of the peasantry he succeeded where a small section of the village community was concerned, but the remainder were seriously offended; for the *mir*, the commune, had formed the framework for their struggles, and its dismantling seemed an act of aggression against them on the part of the Tsar's government.

But the solution to the problem was not to be found exclusively in the rural areas. All the peasants who were obliged to leave their village communities had to be provided for. Emigration to Siberia was one of the solutions conceived by the authorities, and they helped the muzhiks to resettle there. The great majority, however, 'emigrated' to the big industrial cities, to look for work. Thanks to the progress achieved in the economic domain during the previous twenty years, they were able to find it.

The policy initiated by N. Bunge and developed by Witte and his successors had consisted in investing so as to increase the country's potential rather than spending money on increased armed forces that a deficient economy could not sustain. They saw the problem clearly, yet, twenty years later, it had still not been solved, despite the extraordinary development of Russian industry between 1894 and 1913. This failure was the direct cause of the military disasters of 1915, when the army's rear proved incapable of keeping up provision of the supplies needed at the front.

This whole policy of industrialization was based upon the foreign loans that had made possible a 'take-off' in the country's industrial apparatus.[2] Already at this time, however, some people were asking whether such a policy was not turning Russia into a dependency of the Western Powers, a thesis popularized by Lenin. Witte agreed that this was indeed the case, and that it was regrettable, but he said that, for the time being, the country had to go through such a phase. Actually, he claimed, Russia's military power counterbalanced, in a way, the effects of this economic dependence. It is true that, except during the Algeciras conference on Morocco, held soon after Russia's defeat in the Far East, France, her principal creditor (responsible for 27 per cent of the foreign investment), had not in fact dictated the Tsar's foreign policy. France needed the Russian army no less than Russia needed the French investors. This was clearly seen during the

2. French investors showed more interest in what they could receive from a loan at 5 per cent than sympathy with the appeals by Gorky and the Russian socialists warning them against provision of such support for Tsardom. This was why, after 1917, the Russian revolutionaries refused to honour the debts incurred by Nicholas II.

Agadir crisis in 1911, when St Petersburg repaid Paris for its financial help.

In 1913, despite the 'take-off', Russia was still importing between a third and a half of the industrial products she needed. However, she had achieved independence as regards railways and some armaments, even if the results were not yet wholly satisfactory. The real problem was (already) that industrial production, badly balanced, was unable to satisfy the population's requirements in consumer goods. 'Costs of production too high and a bad system of distribution': this diagnosis, which seems topically relevant, dates from 1899.

The masses suffered from this situation, in town and country alike. Prices were such that, given the low level of wages, workers could only just manage to buy the bread they needed. At the same time, speculation and high profits led to the appearance of a class of capitalist magnates whose opulence dazzled the poor. Never had St Petersburg been so glittering – on one bank of the Neva at least – as on the eve of the Romanovs' tercentenary in 1913.

St Petersburg was still living in great style:

At the Hotel Europe the Negro barman had a Kentucky accent. At the Michael Theatre the actresses performed in French. The majestic columns of the imperial palaces proclaimed the genius of their Italian architects. Politicians spent three or four hours at table each day, and the pale rays of the midnight sun, when in June they glided into the shady corners of the gardens, discovered long-haired students discussing with girls the transcendental values of German philosophy. One might doubt the nationality of this city where champagne was ordered in magnums, never in bottles. And yet, there was the monument to Peter the Great, the Emperor in bronze, looking down from his rearing horse upon the grim masses of his city. Opposite, just across the river, lay the fashionable quarters where men played hard, 40,000 of them being registered stock-brokers. Archbishops were not the last to park their vehicles outside the Stock Exchange.

There one might encounter magnates like N. Ryabushinsky who were patrons of the arts and sponsored exhibitions in which Van

Gogh, Rouault and Braque were the stars. There too, among the Russians, were the painters and poets who initiated Futurism; 'an artist in words can be that only if he also can paint', according to N. Kulbin, Mayakovsky, Livshits ...

This little world aimed to live a thousand miles away from the 'committed' writers of the previous decades. Men like Blok, Bely and Balmont – who had evolved — were revolutionaries, they claimed, more by the way they wrote than by the content of their works. When Stravinsky, in 1913, broke with traditional musical forms and created *The Rite of Spring*, and when Diaghilev was transforming the ballet, these poets wanted to do likewise, to 'smash' the language, to invent new words. In a famous 'manifesto' issued in Moscow in 1912, D. Burlyuk, A. Kruchenykh, V. Mayakovsky and V. Khlebnikov declared:

This past of ours has too many people in it ... The Academy and Pushkin are more incomprehensible to us than are Egyptian hieroglyphics. The vessel of modernism throws overboard the Pushkins, Dostoevskys, Tolstoys and their like ... All these Gorkys, Kuprins, Bloks, Sologubs, and so on, what are they after, apart from a *dacha*? In the same way that destiny rewards a mere tailor ...

From the summits of skyscrapers we look down on these less-than-nothing creatures and decree the rights of poets:

To enlarge the vocabulary as much as we want.

To feel an ungovernable hatred for present-day language.

Let good taste and common sense tremble!

Forward, the beauty of the new word

For whom it is enough to say that it is It.

People said that Isadora Duncan was going to dance the Futurists' verses. But Nicholas had long since ceased to attend these performances, which had been marked by the spirit of the age and modernity. All that the avant-garde and the Tsar had in common was dislike of the cinema, and this for different reasons. The avant-garde were hostile to films because they reproduced time-hardened forms: Mayakovsky saw the pre-1914 films as belonging to a realm different from that of art. As for Nicholas, he and his court considered that the melodramas and newsreels produced by film-makers risked perverting the people's morals.[3] The only film the Tsar liked, to the point of even giving its author a ring by way of thanks, was *The Grasshopper and the Ant* (1911), a puppet film by L. Starevich, a brilliant precursor of Walt Disney.

The Futurists wanted to change the words in the language, but Nicholas II would not accept ministers who wanted to change even the accents. As for the people, the workers who lived in the districts on the far bank of the Neva, what they wanted was to learn to read. Was not this divergence a sort of premonition of the break-up that Russia was going to experience?

Alexei Tolstoy recalls a St Petersburg that

lived a restless, cold, satiated, semi-nocturnal life. Phosphorescent, crazy, voluptuous summer nights ... galloping troikas ... duels at daybreak ... ceremonial military parades ... before the terrifying gaze of the Byzantine eyes of an Emperor ... In the last ten years huge enterprises had sprung into being with unbelievable rapidity. Fortunes of millions of roubles appeared as if out of thin air ... People doped themselves with music ... with half-naked women ... with champagne. Gambling clubs, houses of assignation, theatres, picture-houses, amusement parks cropped up like mushrooms ... an epidemic of suicides spread through the city ... Everything was accessible: the women no less than the riches ... Young girls were ashamed of their innocence and married couples of their fidelity to each other. Destruction was considered in good taste and neurosis a sign of subtlety. This was the gospel taught by fashionable authors suddenly emerging from nowhere ...

And to the palace, up the very steps of the imperial throne, came an illiterate peasant with insane eyes and tremendous male vigour: jeering and scoffing, he began to play his infamous tricks, with all Russia as his plaything.

3. 'I consider cinematography an empty, useless and even pernicious diversion. Only an abnormal person could place this sideshow business on a level with art. It is all nonsense, and no importance should be lent to such trash.' This was what Nicholas II thought about the cinema (cf. J. Leyda).

It was said that this man had conquered the Tsaritsa and become the strong man who would take over where Stolypin left off, since neither Kokovtsov nor the aged and feeble Goremykin could do that.

Savvich Pankratov, an eyewitness, tells us what happened on the evening of 1 September 1911, at the Kiev Opera House. The authorities sought to demonstrate their support for the Tsar. Military men in full fig, Polish magnates, Ukrainian nobles, all the smartest people in Russia were present. They had fought each other to get tickets for this special presentation of *The Tale of Tsar Saltan*, by Rimsky-Korsakov. The pit gleamed with white tunics and white dresses. Most of the ministers were present: Kasso, Sukhomlinov, Sabler and, in the front seats, Kokovtsov and Stolypin. At about nine the Tsar arrived, accompanied by two of his daughters, Olga and Tatiana. He occupied a box along with Crown Prince Boris of Bulgaria and the Grand Dukes Andrei Vladimirovich and Sergei Mikhailovich. Pankratov writes:

I had a good view of the ministers, including Stolypin, who were in the front seats to the left, and consequently near the Tsar's box. It was obvious that he was not very interested in the performance; he kept looking to right and left and clearly his mind was elsewhere. In the intervals most people, the Tsar included, went into the foyer, but Stolypin stayed in his seat. A group gathered around him and I recognized his bodyguard, Esaulov.

It was during the final interval that, as I was strolling in the corridor, I heard two sharp reports. I thought something had gone wrong in the electrical system, and was far from suspecting an assassination ... But then I heard: 'Somebody fired ...' One immediately thought of the Tsar ... 'He is alive,' I heard. Tension suddenly increased, we were frozen with anxiety, and I saw Stolypin, white-faced: as they were gently removing his jacket he turned towards the Tsar and, with a last gesture of devotion, made the sign of the cross. I heard shouting in the theatre: 'Kill him, death to him' ... Women's voices ... Meanwhile, the actors, stunned, were still standing there on the stage, their arms stretched towards the imperial box and shouting: 'Hurrah!' The imperial hymn was played three times, and when Nicholas II left the box somebody began singing, on his own, 'God Save the Tsar'. The song was taken up by the entire cast.

In *Istorichesky Vestnik* of October 1911 it was reported that the assassin was named Dmitri Grigorevich Bogrov and that he was Jewish. Pogroms were expected to follow. The man was a revolutionary who had entered the service of the police. The latter knew that he was preparing to kill Stolypin, and the chief of police himself had provided him with a ticket for the performance. He was hanged before the senator appointed to investigate the affair could reach Kiev.

At the time it was said, wrongly, that the Tsar did not even go to the bedside of the mortally wounded minister, who died three or four days after the shooting. Nicholas told his mother about it all in a letter of 10 September 1911:

On ... 27 August we went to Kiev ... On 1 September, at the theatre, this dastardly attempt was made on Stolypin's life. Olga and Tatiana were with me at the time. During the second interval, we had just left the box, as it was so hot, when we heard two sounds, as if something had been dropped. I thought an opera-glass might have fallen on somebody's head, and ran back into the box to look.

To the right I saw a group of officers and other people. They seemed to be dragging someone along: women were shrieking, and directly in front of me in the stalls Stolypin was standing. He slowly turned his face towards us and, with his left hand, made the sign of the cross in the air.

Only then did I notice that he was very pale and that his right hand and uniform were bloodstained. He slowly sank into his chair and began to unbutton his tunic. Freedericks and Professor Rein helped him. Olga and Tatiana came back into the box and saw what happened. While Stolypin was being helped out of the theatre there was a great noise in the corridor near our box: people were trying to lynch the assassin. I am sorry to say the police rescued him from the crowd and took him to an isolated room for his first examination. He had been badly manhandled, however, and two of his teeth had been knocked out. Then the theatre filled up again, the national anthem was sung, and I left with the girls at eleven. You can imagine with what emotions!

Alix knew nothing about it till I told her. She took the news rather calmly. Tatiana was very much upset, she cried a lot and they both slept badly.

Poor Stolypin passed a very bad night and had to have several injections of morphine. The next day, 2 September, there was a magnificent review of troops at the place where the manoeuvres had ended, 50 *versts* from Kiev ...

I returned to Kiev on the evening of 3 September, called at the nursing-home where Stolypin was lying, and met his wife, who would not let me see him. On the 4th I went to the first school to be founded at Kiev, which was celebrating its 100th anniversary.

Nicholas spent more and more of his time at Peterhof. He would have preferred to live at Livadia, in the Crimea, in the new palace he had had built there. In that place he relaxed, played tennis, bathed, took a real holiday. Since Moscow no longer meant for him anything but bad memories, Tsarskoe Selo had become his favourite residence. That little Versailles, close to the earth, was to his liking. There he was on good terms with the palace commandant, responsible for his security. Count Freedericks held this post, and kept it till 1917. 'Elegant, impeccable, he was much admired when, on his regiment's festival day, he rode by, wearing a cuirass and a golden helmet surmounted by an eagle, on a black Arab horse.' Freedericks was in charge of His Majesty's personal escort, made up of Caucasian horsemen, the finest of all the detachments entrusted with the Emperor's protection. Also under his command was the battalion composed of specially chosen men that had been created after the assassination of 1881. The palace police were likewise subject to him; they patrolled the surroundings of the imperial residences at Tsarskoe Selo and elsewhere.

For some years now the Tsar had been living more and more shut away, going less and less often to the theatre, and staying at Tsarskoe Selo or Livadia in his little private world, with his wife (whom he worshipped), his four daughters and Alexei.

Olga was the eldest of the girls. Intelligent and lively, she was simple and direct in her ways, a typical Russian girl, sensitive and affectionate, upright and determined. She had refused categorically to marry Carol, the Crown Prince of Romania, because, she said, she was Russian and wanted to stay in Russia. Tatiana, the second daughter, had a dignified air and outshone her sisters in elegance, slimness and beauty. Though outwardly cheerful she was reserved and pious, like her elder sister. Maria, the third, was Nicholas's favourite. Coquettish and with winning ways, she thought only of love, marriage and children. Like her father, she was patient and mild. Anastasia, the youngest, was the clown of the family – she should have been a boy, people said – and tossed paper pellets over the soldiers when mobilization came. She was the very opposite of her father, who secretly admired this reckless little creature.

Alexei was a charming lad, intelligent and fond of jokes. But he was not joking when, aged six, he said to the chairman of the Council of Ministers: 'When the heir to Russia's throne enters a room, people must get up.' Despite his illness, nothing serious had happened during his infancy. The four sisters (who were very close and signed their letters with 'OTMA', the initials of their Christian names) looked after Alexei and watched over him. They knew that he must not cut or bruise himself. But the little boy was wild, like Anastasia, and the family had to be constantly alert to what he was up to. The first serious accident occurred at Spala, where the Tsar had gone to hunt bison. Alexei, who was not allowed to ride a horse, was allowed to go out in a boat on one of the lakes in the forest of Bialowieza. He had an awkward fall and the resultant bleeding took a long time to stop.

The health of the heir to the throne now became an obsession with his family. A haemorrhage could prove fatal – and Alexandra bore the responsibility for that, a thought that added to the anxiety she felt as a mother. Given these conditions, we can appreciate the passionate attachment she formed for Rasputin when that man, introduced to her by one of her ladies, Anna Vyrubova, showed himself able on more than one occasion, by means of hypnotism, to relieve the child's suffering. For the Tsaritsa, who was already very credulous, this amounted to a miracle. Her mysticism increased, and Rasputin's power grew greater with every passing day.

A little group of mystics, preachers and hypnotists began to gather round the Tsaritsa. There was first a certain Philippe, then his pupil Papus and soon Rasputin, who was recommended by Feofan, the Rector of the Theological Academy of St Petersburg. Appointed the Tsar's *lampadnik*, Rasputin was responsible for seeing to the little lamps that had to be kept continuously lit before the holy icons. He was thus at the palace all the time, and met the Tsar, who was very fond of icons and gave Rasputin the task of lighting up a very precious collection.

At the same time Rasputin expanded his connections by organizing mystical gatherings like those held as part of the ritual of the Khlystovskie Korabli, one of the many sects that had developed in the shadow of Orthodoxy. These 'sectaries', known as *khlysty*, or 'flagellants', called themselves 'worshippers of the living God', and personified the deity in a man who was His provisional representative, so that they had a whole series of 'Christs'. They rejected all written knowledge and had symbolically cast the Scriptures, which they held to have been falsified, into the Volga. The success enjoyed by the *khlysty* owed more to their rites than to their beliefs. Like the early Christians they assembled secretly, and when together they sang and danced until they were exhausted, for they believed that the way to ecstasy lay through the senses. Men and women alike flogged themselves in a frenzied whirlwind, arriving at last at a state of fervour (*radenie*) that, around a vat of boiling water, sometimes culminated in 'the collective sin', the sin of the flesh, when beating gave way to religious lust, for this sin of the flesh was regarded as a means of overcoming spiritual pride.

These practices, originating in the depths of the countryside, had reached the capital and become widespread among the aristocracy in the form of a branch of the *khlysty* known as the *skakuny*, because instead of dancing its adepts jumped up and down before giving themselves over to the pleasures of the love of Christ. Neither the Tsaritsa nor the Tsar took part in these gatherings, but they became notorious at court. In organizing them at Tsarskoe Selo or St Petersburg, Rasputin was therefore not necessarily doing anything out of the ordinary, even if, according to an analysis by Bonch-Bruevich, a specialist in the affairs of the sects, this man was not a *khlyst* or any other sort of sectary, but perfectly Orthodox.

The gatherings arranged by the *starets* (the term used for holy men who were not in orders) acquired a new importance. Entrusted by the Duma with the task of reporting on Rasputin's activities, Speaker Rodzianko recalled the man's first, still modest, success and how he went about increasing his clientele. Having learned that he had influence with the monarch, a provincial lady came to St Petersburg and obtained an audience with Rasputin. She asked him to put in a word that would ensure her husband got promotion.

He scolded me severely and said: 'Aren't you ashamed? Come to me to repent, but with your shoulders bare; otherwise, don't come . . .' He pierced me with his gaze and took liberties with me, so that I left, indignant, deciding not to approach him again. However, anxious and worried, disturbed, too – fascinated, in short – I obtained a low-cut dress and, palefaced, went to see him. He received me and gazed fixedly at me once more, staying at a distance from me . . . A few days later my husband got his promotion.

Accused of immorality by Bishop Hermogen, who realized the evil consequences that could result from his presence at court, Rasputin was summoned before the Bishop. Present also were the monk Iliodor, an army officer named Rodionov, who was also a writer, a lay brother and a pilgrim named Mitya. 'I call upon you', said Hermogen, 'never to set foot again in the Tsar's court.'

Rasputin's answers to the indignant bishop were insolent, Rodionov tells us, and a violent scene followed, in the course of which Rasputin, after abusing Mgr Hermogen in vulgar language, flatly refused to submit to his command and threatened to 'make short work of him'. At this Bishop Hermogen, losing his self-control, exclaimed: 'So, you dirty scoundrel, you refuse to submit to my episcopal command and dare to threaten me. Then know that I, as a bishop, anathematize you!' Rasputin, as if possessed, clenched his fists and flung himself on the prelate: his face, said Rodionov, losing every trace of humanity. Afraid lest, in his access of fury, Rasputin should murder the bishop, Rodionov drew his sword and, with the others, hastened to the rescue. With difficulty they managed to drag Rasputin away. His bodily strength was such that he managed to wrench himself from their grasp and take to his heels. He was, however, overtaken by the lay brother and the pilgrim Mitya and roughly handled. Still he managed to escape and ran out into the street shouting: 'You wait a bit, I'll pay you out.'

'Where are we going?' cried V. Purishkevich, a deputy of the extreme Right, addressing Rodzianko:

They want to destroy our last rampart, the Holy Orthodox Church. We have already seen a revolution which encroached upon the supreme power ... Fortunately, it failed. The army stayed loyal. Now, to crown our sorrows, the forces of darkness are attacking Holy Russia's last hope, her church ... And the worst aspect of the matter is that these sorrows appear to be descending on us from the height of the imperial throne. A vile impostor, a *khlyst*, an illiterate muzhik, dares to mock our bishops ... I should like to sacrifice myself in order to kill this scoundrel.

The scandal grew, and rumour with it. In 1911 the newspaper *Golos Moskvy* published a letter to the editor headed 'the cry of an Orthodox parishioner' and signed by the managing editor of the periodical the *Religious and Philosophical Library*, Michael Novoselov:

Quousque tandem? Such is the cry of indignation that bursts from the lips of all Orthodox men and women against that cunning conspirator against our Holy Church, that fornicator of souls and bodies – Gregory Rasputin, who acts under the holy cover of that Church. *Quousque?* Such are the words that, in anguish and bitterness of spirit, the sons of the Orthodox Church are compelled to address to the Synod, in view of the unheard-of tolerance exhibited towards the said Gregory Rasputin by the highest dignitaries of the Church.

How are we to explain the silence of the Bishops who are quite well aware of the behaviour of this trickster? Why are the guardians of the Faith silent when, in the letters they send me, they describe this propagator of false doctrine a pseudo-sectary, a sex maniac and a charlatan? What becomes of the Holiness of the Holy Synod if, through indifference or cowardice, it does nothing to maintain and safeguard the purity of the Christian religion, but, instead, allows a so-called sectary to cover all his swindles with a veil of sanctity? Where is its power, if it is unwilling to make any attempt to protect the Church from defilement by the proximity of this heretic?

The confiscation of Novoselov's appeal and the interpellation in the Duma regarding this confiscation provided glaring confirmation of the rumours about Rasputin's influence and behaviour at the Tsar's court. 'Nobody can have doubts any longer concerning the truth of everything that has been said about this person,' Rodzianko commented. As publicity developed around these matters and an impressive number of letters were sent to Rodzianko by mothers whose daughters had been abused by Rasputin, the Tsar eventually agreed to receive the tiresome Speaker.

'Have you so much as met him?' Nicholas demanded.

'I have always avoided meeting him, Sire. An adventurer like that fills me with repulsion.'

'You are wrong in your opinion of him; he is just a good, simple-minded, religious Russian. When troubled or assailed by doubts, I like to have a talk with him, and invariably feel at peace with myself afterwards.'

Rodzianko summarized for the Tsar the finding of his report, mentioned the incident with Hermogen and read out the article from *Golos Moskvy*.

'But why do they keep on accusing Rasputin, why do they find him dangerous?'

'A number of bishops have been removed from office through him. That has worried people.'

'I think Hermogen is a good man. He will soon be permitted to return. Still, I could not allow him to remain unpunished for his flagrant disobedience to my imperial order ... And what proof have you that Gregory is a *khlyst*?'

'The police discovered that he went to the baths with women.'

Nicholas II replied: 'What of that? It is merely a custom among the common people.'

Rodzianko mentioned the names of several women in high society who had been seduced by Rasputin, including one of the maids of the imperial children, Engineer Laktin's wife, who had subsequently gone mad and been put away.

The Tsar seemed impressed. He smoked cigarette after cigarette, without even finishing them.

'Do you believe me, Sire?' Rodzianko asked.

'Yes, I believe you,' the Tsar replied, obviously disturbed.

'Will you authorize me to tell everyone that he will not return to court?'

The Tsar hesitated for a moment, then replied: 'No, I cannot promise you that. Nevertheless, I fully believe all you have told me.'

The family circle grew narrower and narrower. Anna Vyrubova had become the Tsaritsa's confidante. There was even a rumour that she had improper relations with Alexandra. She was said to be stupid and a scandalmonger. The diary attributed to her throws light on the relations between her, Rasputin and the Tsaritsa on the one hand and Nicholas on the other. All three strengthened the Tsar in his notion that the Duma was the enemy of Tsardom and that for him to agree that the government should be responsible to this assembly would be an act of surrender. The presence of Rasputin, who for the Tsar embodied the church, the muzhik and Holy Russia, was a reassurance to this distressed little group of people who could find no support either in the city or at court, where the Grand Dukes and Princes were muttering against their incapable Tsar.

Nicholas's relations with his kinsfolk grew even worse. In the first place, he had continually to interfere to prevent a row between his mother and Alexandra. The two women could not get on together. The Dowager Empress liked fun, she was gay and sensual, intelligent and somewhat frivolous. Alexandra was austere and fond of solemn ceremonies. Whenever it emerged, the discord between the two women embarrassed Nicholas. The Dowager Empress advised her son to surround himself with intelligent men like Witte and Stolypin, while the young Empress recommended more respectful types like Goremykin and Stürmer. Their only point of agreement was in loathing Willy – Wilhelm II — and preferring Uncle Bertie – Edward VII.

Then there were the clashes between Nicholas and his uncles. Although, as a rule, the Tsar deliberately sided with the Niko-laeviches against the Mikhailoviches, Sergei Mikhailovich had been one of the few whose advice he listened to, while neither Nicholas nor Alexandra could stand any longer Grand Duke Nicholas Niko-laevich, who was virtually commander-in-chief.

And now there was the business of Cyril, Uncle Vladimir's son, which had a marked effect on Nicholas, who was devoted to pro-priety in the imperial family. Though Cyril had no right to remarry, he disobeyed the accepted usage. Nicholas heard about this in the midst of the 1905 Revolution. On 5 October he wrote to his mother:

Dearest Mother, this week there has been a family drama, with the unfortunate marriage of Cyril [to Grand Duchess Victoria Fyodorovna]. You will remember the sanctions to which he was subject if he married. I spoke to him about them. First, dismissal from the army. Second, ban on entry into Russia. Third, loss of all income from his apanages. Fourth, deprivation of the title of Grand Duke. Well, on 25 September he got married ... I learned the news from Nicky [Prince Nicholas of Greece], who told me while we were out hunting that Cyril was arriving next day! I must admit that I was extremely annoyed by this impertinence. An impertinence it was, because he knew quite well that, after his marriage, he had no right to return. In order to stop Cyril coming to our house I summoned Freedericks and ordered him to go to Tsarskoe, remind Cyril of the four sanctions and tell him how angry I was with him ... Next day, as though it had been deliberately arranged, we had to receive Friedrich-Leopold [of Prussia], a bird of ill omen. He lunched with us, along with Uncle Vladimir [Cyril's father] ... who took his son's side, and, at his request, I agreed that he leave the army ... After that I heard no more of Cyril, apart from a letter from Nicky asking me to mitigate his punishment.

The order of the day for the navy appeared all right [Cyril was commander of the Marine Guards] but they did not manage to produce the document which set out his deprivation of the title of Grand Duke. It was the first time this had happened.

At the same time I was seized by doubt. Was it a good idea to punish publicly the same person in several ways at once, and at a time when people are looking unfavourably on our family in general?

After thinking this over at length, until I had a headache, I decided to take the opportunity of the birthday of your dear little grandson [the Tsarevich Alexei] to telegraph to Uncle Vladimir that I was giving back to Cyril the title he had lost ... I really feel that I have at last rid myself of a mountain that was weighing on my shoulders ...

Excuse me, dearest Mother, for filling my letter with this one subject, but I wanted you to learn the whole truth from me. May Christ be with you. Your Nicky, who loves you with all his heart.

'The men around the throne are parasites,' commented the wife of General Bogdanovich. 'Grand Duke Alexander Mikhailovich leads a life of luxury ... Under Astashev, its captain, debauchery prevailed on the imperial yacht ... Astashev diverted the Tsar by showing him his collection of pornographic postcards.' Though Astashev's dishonesty had been exposed,

nothing stops the Tsar from showing him sympathy and providing him with well-paid jobs. Subjects like him contribute to the downfall of monarchies. The Tsar is on bad terms with Grand Duke Nicholas Nikolaevich ... In Paris, Grand Duke Boris Vladimirovich went with prostitutes during mid-Lent. They wound paper streamers round him ... In Monaco Grand Duchess Maria Pavlovna has lost a fortune at roulette. Grand Duke Alexei Alexandrovich is there too, with La Baletta (his mistress), who costs more than our defeat at Tsushima.

Nicholas II knew all about all that. But whenever the name of a member of his family or that of one of his protégés was uttered, his face became inscrutable. He looked fixedly out of the window, ending the audience. Yet he was aware that he could not count on his kinsfolk for help. When, in August 1906, he assembled the Grand Dukes in order to take up the question of their handing over their land to the peasants, Grand Duke Vladimir Alexandrovich told the Tsar that this proposal made him think of anarchists shouting: 'Hands up!' They were all against this idea, which Stolypin had put up to the Tsar, even though it would have brought them 6 million roubles in compensation.

If there was one relative who shared Nicholas's troubles with him, it was Willy. During 1905 and 1906 he bombarded the Tsar with letters and telegrams. Nicholas read them, we know, but he rarely replied to them. Wilhelm II mingled information ('The Japs have just ordered four line-of-battle-ships from England') with advice, relating both to the war ('It's unpopular, you should end it') and to domestic policy:

What terrible tidings have come from Moscow [that is, the assassination of Grand Duke Sergei] ... I cannot believe that these demons have risen from the ranks of your Muscovite subjects, they were probably foreigners from Geneva. For the great bulk of your people still place their faith in their 'Little Father' the Tsar and worship his hallowed person ...

The example of Nicholas I has often been quoted, who quelled a very serious rebellion by personally riding into their midst, his child in his arms ... A word from such a position and such an 'entourage' would have awed and calmed the masses and sounded far away over their heads into the furthest corner of the realm, surely defeating the agitators. These are still more or less said to be in command of the masses because such a word has not yet been spoken by the Ruler ... Many reforms have been begun and new laws are being discussed in batches, but, curiously enough, the people generally say: 'This is by Witte ... that is Pobed's [Pobedonostsev's] idea' ... In an autocratic regime ... it must be *the Ruler himself* who gives out the password ... It seems that everybody is expecting something of this sort by way of an act of will by the Tsar personally.

To this solicitude Nicholas was not wholly indifferent; for what Willy told him fitted in with what he himself thought. He felt that he had been let down by France since she entered into the Entente Cordiale with Britain, Japan's ally.

When Wilhelm suggested a personal meeting, on 6 July, he agreed, and proposed that the place be the island of Björkö. Family troubles and international affairs delayed the encounter. Eventually, however, 'we cast anchor off the island of Kavitsa ... We awaited the arrival of the *Hohenzollern*, which was two and a half hours late. The yacht came up just as we were about to have supper. Wilhelm came over to my yacht in an excellent mood ... He took Misha and me back to his yacht and gave us supper.' In his diary, just this remark: 'I returned with a very good impression from the time spent with Wilhelm ... I am glad to see the children again, and not to see the ministers.'

In fact, behind his ministers' backs, Nicholas had signed with his cousin the treaty of Björkö which was a veritable *renversement des alliances*: 'If one of the two states is attacked, its ally undertakes to come to its aid.' The Franco-Russian alliance lost its validity, since Russia had now made the same commitment to Germany. Article 4 proposed that France be requested to associate herself with this treaty.

Wilhelm was delighted. Nicholas, however, was a little worried. Admiral Birilev, the Minister of Marine, told Witte that the Tsar summoned him to his stateroom and asked him, point-blank: 'Have you faith in me, Alexei Alexeyevich?'

'Naturally,' said the Admiral, 'there could be but one answer.'

'In that case,' His Majesty went on, 'sign this paper. It is signed, as you see, by the German Emperor and myself, and countersigned, on Germany's side, by the proper official. Now the German Emperor wants it to be countersigned by one of my ministers.' Birilev applied his signature to the paper.

The treaty of Björkö was to come into force only after peace had been concluded between Russia and Japan. On Russia's side the negotiations were conducted by Witte. In a coded telegram Nicholas told Witte on what conditions he would make peace:

For the last three months I have thought much about the question of peace

Alexander III and his family.

Nicholas II and the Prince of Wales (later, George V) in 1909.

Nicholas II, a Maharaja and Prince George of Greece.

Empress Alexandra Fyodorovna prepares for her first ball.

Nicholas II with the Tsarevich and the Grand Duchesses Olga, Tatiana, Maria and Anastasia.

The coronation: Nicholas II sets the crown on the Empress's head, May 1895.

Ceremonial entry by Nicholas II into Balmoral, 1896.

Members of the government and of the *zemstvos* on the fiftieth anniversary of the creation of the *zemstvos*.

Their Majesties leaving the Trinity Cathedral of the Ipatiev Monastery.

P. Stolypin.

V. Plehve.

S. Witte.

D. Sipyagin.

Nicholas II with his officers, 1914.

Opening of the First Duma, 27 April 1905.

Execution of revolutionaries.

ВЫСОЧАЙШІЙ МАНИФЕСТЪ.

БОЖІЕЮ МИЛОСТІЮ.

МЫ, НИКОЛАЙ ВТОРЫЙ.

ИМПЕРАТОРЪ И САМОДЕРЖЕЦЪ ВСЕРОССІЙСКІЙ.

ЦАРЬ ПОЛЬСКІЙ, ВЕЛИКІЙ КНЯЗЬ ФИНЛЯНДСКІЙ,

и прочая, и прочая, и прочая.

Leaflet: 'The Bloody Tsar', 1905.

or war ... There are two points upon which every good Russian agrees to continue this fight to the end if Japan insists on them – not an inch of our territory, not one rouble of war indemnity ... Nothing ... will induce me to consent to these two demands. Therefore there is no hope for peace for the present. You must know how I hate bloodshed, but still it is preferable to an ignominious peace.

In fact, thanks to the intercession of the Americans, who were already hostile to Japan, and thanks to Witte's skill, Japan, while acquiring, under the treaty of Portsmouth, the Liaotung Peninsula, a protectorate over Korea and the southern half of Sakhalin, did not get an inch of strictly Russian territory (unless one includes south Sakhalin in that category) and she agreed not to insist on a war indemnity.

Witte was able to return home satisfied. So was Wilhelm, since the Björkö treaty now came into force. He made Witte a count and gave him the Order of the Black Eagle. Nicholas looked on this action as extremely tactless: Witte had been decorated by a German before receiving a reward from his own Tsar. And he resented Witte's having succeeded in a situation of which he himself had despaired. Decidedly, Witte could be tolerated no longer.

For his part, Wilhelm rejoiced, and sent Nicholas another message, reminding him how France had let him down:

I got some French newspapers in which I read a résumé of the Brest *fêtes* [the visit of the British Atlantic Fleet to Brest and the return visit to Cowes by the French Northern Squadron]. Il y a 12 ans nous avions Toulon et Cronstadt: c'etait le mariage d'amour. Comme chez tous les mariages d'amour est survenu un désillusionnement général, surtout après la guerre de 1904–05. Maintenant nous avons Brest et Cowes: c'est le mariage d'affaires, et comme chez tous les mariages d'affaires il en resultera un mariage de raison! [Twelve years ago we had Toulon and Kronstadt (the reciprocal visits of the Russian and French fleets): that was a love marriage. As with all marriages for love, all-round disenchantment ensued, especially after the war of 1904–5. Now we have Brest and Cowes: this is a business marriage, and, as with all such, the result will be a marriage of convenience!] I think that is really cool! ... It will do the French a world of good if you draw the reins a little tighter. Their ten milliards of francs they placed in Russia of course hinder them from quite falling off ... To use the metaphor of 'marriage' again, 'Marianne' must remember that she is wedded to you and that she is obliged to lie in bed with you and eventually [*sic*] to give a hug or a kiss now and then to me, but not to sneak into the bed of the ever-intriguing *touche-à-tout* [busybody] on the Island.

Nicholas was certainly put out by the diplomatic assistance that Britain had given Japan, and even more by the extreme discretion shown by his French allies in expressing their sympathy with him during the war. The Björkö treaty gave expression to a twofold resentment and also to a resurgence of the 'diplomacy of monarchs'.

When he got back to St Petersburg, however, Count Lamsdorf, the Minister of Foreign Affairs, together with Witte and Grand Duke Alexander, made him realize that it was madness to weaken in this way the alliance with the Republic, because France was 'the moneybox'.

Nicholas appreciated the force of their arguments; the treaty was not ratified, and it came to nothing. From this ensued a twofold bitterness. Wilhelm bore a grudge against Nicholas for having broken his promise. He saw this as evidence that the Tsar was not master in his own country. Nicholas, realizing this, felt humiliated before the Kaiser. Furthermore, he felt that at Björkö he had let himself be got the better of. And his ministers had told him where he had gone wrong: another wound that did not heal.

The two Emperors met several times again before 1914, but the old feeling between them had gone. Besides, after the death of Christian IX in 1906, the family reunions in Copenhagen were no longer what they had been in the good old days. The diplomacy of statesmen resumed now its ascendancy over the personal relations which briefly resurfaced at Björkö. The diplomacy of courts was dead with Christian IX. Christian X tried to revive it, but without success. In 1915, in the midst of the European war, he wrote to

Nicholas urging him to contact his cousin Willy again. Again, on the Kaiser's initiative, Christian X proposed to Nicholas that he send an emissary to Copenhagen. However, the Tsar rebuffed these approaches. At the end of 1916 there was some question of a visit to Copenhagen by Protopopov, but it is not clear what its purpose was, or whether it ever took place.

In the meantime, relations between Germany and Russia had seriously worsened, as a result of the crises in the Balkans, which were to lead to the first world-wide conflict.

Up to now Russia's policy had been based upon a particularly close alliance with France, friendly relations with Germany, an agreement with Austria-Hungary to preserve the status quo in the Balkans and a degree of co-operation with Britain that was upset by the Anglo-Japanese accord of 1902 and the Far Eastern crisis.

What disturbed this balance was the worsening of the naval rivalry between Britain and Germany. This resulted from the Kaiser's challenge to British dominion of the seas, which compelled the British government to keep up the 'two-power standard', whereby the Home Fleet had to be as strong as the combined fleets of any two foreign powers that might oppose it. Britain now, under French influence, drew closer to Russia; and, on British initiative, negotiations began between Arthur Nicolson and A. Izvolsky.

The Anglo-Russian agreement concluded in August 1907 amounted to a partition of Persia into zones of influence, while Britain retained her position in Afghanistan but gave ground in Tibet. In the course of these negotiations Nicholas met Wilhelm with a view to putting an end to the coolness that had followed the non-fulfilment of the Björkö treaty. The new treaty with the British did not help. So as to make its meaning less obvious, France and Russia signed a treaty with Japan, but this did not deceive Chancellor Bülow, who saw it as a bad omen for the future of peace in Europe – in spite of Nicholas's repeated assurances to Germany of his unfailing friendship.

It was certainly the case that the Tsar did not wish to put the Kaiser's back up, and Stolypin expressed his ideas well when he said that the Russian eagle had two heads, one turned to the East and the other towards Constantinople. The alliance with Britain might help Russia to open the Straits.

Edward VII's visit to Reval in 1908, soon followed by the arrival of President Fallières, set the seal on this Triple Entente, which was further strengthened by the Tsar's visit to Britain and France in the following year. There was a sort of paradox here, which British Liberals and Edouard Vaillant in France did not fail to emphasize; the 'Stolypin reaction' was at its height, and the Western democracies, drawing close to the autocracy, chastely averted their eyes.

To be sure, the opposition in the Third Duma (the Cadets) enthusiastically applauded the treaties made with countries that had parliamentary regimes: their practices might prove contagious. Moreover, as the historian Milyukov observed, these treaties made possible a strengthening of Russia's position in relation to the place where lay the roots of Russian culture, namely, Constantinople.

A kind of imperialism was thus still inspiring Tsarist policy. Dietrich Geyer has shown that this was not so much an expression of the desires of Russian capitalists, even though they did pursue expansionist aims in those isolated regions where they had little to fear from British or German competition. The Russian imperialism of that period was mainly motivated by desire to make up, externally, for the difficulties encountered by the regime on the home front, on both the economic and the political planes. But Nicholas suffered two more humiliations in foreign affairs, serious ones this time. First, in the Balkans, where Tsarist diplomacy was supporting the efforts of Peter I Karageorgevich, King of Serbia, to unite all the southern Slavs in a single 'Yugoslav' state.

The crisis that suddenly broke in 1908 was the first move in the game of chess in the Balkans that would end in the Great War. Austro-Russian relations in that region had been frozen since 1897, but they worsened when Vienna sought to construct a railway through the Sanjak of Novi-Bazar to Constantinople, as a way of keeping closer check on Serbia. Profiting by the domestic crisis in Turkey, Vienna annexed Bosnia-Herzegovina at the very moment when Aerenthal, Franz-Josef's minister, was negotiating with Izvolsky about compensation to be given to Russia. The Tsar looked as though he had been made a fool of when, in reply to Serbia's protest against the annexation, Wilhelm sent that country an ultimatum. As for the conference that Izvolsky demanded in order to settle the question, it never met.

Nicholas II felt this rebuff all the more keenly because foreign affairs were still, and more than ever, a field he reserved to himself. Even Prime Minister Stolypin had not been informed of the negotiation between Aerenthal and Izvolsky at Buchlau. It was the Tsar himself who ordered Izvolsky not to talk about it, Russia's Foreign Minister was later to tell Kokovtsov, when the latter was Stolypin's Minister of Finance. Russia's military impotence three years after the defeat in the Far East made compromise hard for her to accept, because it would be seen as yet another retreat.

These were the circumstances that made useless the great family reunion at which, as in happier times, the members of the Hesse, Hohenzollern and Romanov families assembled together. The garden party held at Potsdam in 1910 led to nothing. The climate no longer prevailed for that sort of diplomacy to flourish.

The 'BBB' (Berlin–Baghdad–Bahn) project and Germany's unconditional support for the Dual Monarchy in the Balkans showed how deeply interested the Central Powers were in those Straits on which Tsarist diplomacy pursued designs once more, overestimating the value in that direction of the British and French alliances. Operating on his own, Izvolsky's successor Sazonov succeeded in bringing the little Balkan states together, in the hope of blocking the *Drang nach Osten* of the Austrians and Germans. He succeeded – except that the Serbs, Bulgars and Greeks, instead of joining against Austria, launched an attack on Turkey which proved victorious. This obliged the Russians to stop the Bulgars from occupying Constantinople and then to oppose the Serbs' demands, which resulted in a second Balkan war, this time between the victors in the first. Russian diplomacy was utterly discredited. It was given the *coup de grâce* when the Sultan called in the German General Liman von Sanders to reorganize his armies and take command of them. In face of a protest from the Tsar, Liman von Sanders' post was given a more discreet title.

But the damage was done, and the theme of 'the struggle of

the Slavs against the Germans' re-emerged. The Russian press, especially, burst out into a Germanophobia that had more than one target: was not the Tsaritsa of German origin?

'Willy' had thus, on two occasions at least, frustrated the plans of 'Nicky', and embittered relations between Germany and Russia. It was no good for P. Durnovo to demonstrate, in a report to the Tsar, that a war with Germany and Austria, even if victorious, would only endow Russia with new territories that were 'useless and dangerous', Poznania and Galicia, acquisition of which would render the Polish question even more insoluble, while at the same time bringing that of Ukrainian autonomy higher up on the agenda. The Triple Entente was strengthened, even though 'the alliance with Britain is of no benefit to us' and the alliance with France risked involving Russia in a war with Germany, Austria and Turkey that might end in disaster for her.

Nicholas, however, still peace-loving, was unable to conceive for one moment that the assassination in Sarajevo could draw Russia into war. 'It's just another Balkan crisis.' He waited till the Austrians sent their ultimatum to Belgrade before saying that, this time, he would stand by Serbia. Nicholas called on Britain to take up a position, to summon the Hague Court or to demand an international conference. But the machinery had been set in motion. After the declaration of war on Serbia by Austria, the Tsar ordered partial mobilization – first against Austria only, then with general application. He made no reply to the ultimatum from Wilhelm demanding that he suspend these war preparations.

Nicholas felt that these were decisive moments. He had sent a telegram to Wilhelm: 'The right thing to do would be to submit the dispute between Austria and Serbia to the Hague Court. I rely on your wisdom and your friendship.'

A few days later, Nicholas said to Sazonov, as the latter recalls:

He is asking the impossible. He has forgotten, or does not wish to remember, that the Austrian mobilization began sooner than the Russian, and now asks us to stop ours without saying a word about the Austrian.

You know I have already suppressed one mobilization decree and then consented only to a partial one. If I agreed to Germany's demands now, we should find ourselves unarmed against the Austrian army, which is mobilized already. It would be madness ...

[I] sat opposite the Tsar, watching him intently. He was pale and his expression betrayed a terrible inner struggle. I was almost as agitated as he. The fate of Russia and of the Russian people depended upon his decision ... We were in an impasse ...

At last the Tsar said, speaking as it were with difficulty: 'You are right. There is nothing left us but to get ready for an attack upon us. Give then the Chief of the General Staff my order for mobilization.'

On 2 August 1914, in the afternoon, the Tsar issued a manifesto to his people. The only foreigner allowed to be present on this solemn occasion, being the representative of Russia's ally (Britain did not declare war until two days later), was Maurice Paléologue, the French Ambassador. Here is his account of it.

It was a majestic spectacle. Five or six thousand people were assembled in the huge St George's gallery which runs along the Neva quay. The whole court was in full dress, and all the officers of the garrison were in field dress. In the centre of the room an altar was placed and on it was the miraculous icon of the Virgin of Kazan, brought from the national sanctuary on the Nevsky Prospekt, which had to do without it for a few hours ...

In a tense, religious silence the imperial cortège crossed the gallery and took up station on the left of the altar. The Tsar asked me to stand opposite him as he desired, so he said, to do public homage in this way to the loyalty of the French ally.

Mass began at once, to the accompaniment of the noble and pathetic chants of the Orthodox liturgy. Nicholas II prayed with a holy fervour that gave his pale face a movingly mystical expression. The Tsaritsa, Alexandra Fyodorovna, stood by him, gazing fixedly, her chest thrust forward, head high, lips crimson, eyes glassy. Every now and then she closed her eyes and then her livid face reminded one of a death mask.

After the final prayer the court chaplain read the Tsar's manifesto to his people ... Then the Tsar went up to the altar and raised his right hand towards the Gospel held out to him. He was even more grave and composed, as if he were about to receive the sacrament. In a slow, low voice ... he made the following declaration: 'I solemnly swear that I will never make peace so long as one of the enemy is on the soil of the fatherland.'

A wild outburst of cheering was the answer to this declaration, which was copied from the oath taken by the Emperor Alexander I in 1812. For nearly ten minutes there was a frantic tumult in the gallery and it was soon intensified by the cheers of the crowd massed along the Neva.

Germany, followed by Austria, at once declared war on Russia. Germany then declared war on France, after which Britain joined her two allies, as had been promised.

3. Defeat: The Tsar Broken

In France, when war threatens and the country is in danger, it is the sound of the bugle and the sight of the tricolour that summon up our energy and lift up our hearts. In Russia, it is the bells. By which I mean the profound resonance of manifestations of religious faith, their identification with the nation and the fatherland. There is no Russian hymn in honour of the country that does not also hail the Orthodox faith. During the wars against the infidel Turk the bells were rung, and they rang again in 1914.

In 1905 they had not been heard. True, there had been some patriotic demonstrations, but these involved students and soldiers only. The Tsar had not made a real appeal to his people, for neither he nor they, at the start, considered a brush with Japan to be a war. In 1914, though, when the Germans growled, the Tsar called upon God for help.

Was not the church there too, the church that, every year, addressed those who do not believe that the Orthodox monarchs have been raised to the throne by virtue of a special grace of God – nor that, at the moment the sacred oil is laid on them, the gifts of the Holy Ghost are infused in them anent the accomplishment of their exalted mission? Such persons were to be treated like atheists and heresiarchs: 'Anathema! Anathema! Anathema!'

Nicholas II was pious but not really mystical in his beliefs. When the revolutionary movement surged up, when his armies and fleets were defeated in 1905, when illness inexorably gripped his son Alexei, religion was for him a protection and a refuge. In 1914 it was a support.

The announcement of the declaration of war was to give the Tsar his greatest joy. He went to St Petersburg with his whole family, to attend a solemn service of thanksgiving in the presence of all the court. After the mass the Archdeacon read out the declaration of war. In accordance with the time-honoured formula it called on the soldiers to go into battle 'with sword in hand and cross on heart'. Everyone present rushed towards the sovereign, to kiss his hand. Then Nicholas and Alexandra appeared on the balcony.

The Pathé newsreel shows us the unprecedented scene of thousands of Russians on their knees and then rising to their feet with cries of greeting to the Tsar, waving banners inscribed with the words: 'Help our little brother Serbia.' The Tsar and the Tsaritsa seem surprised, astonished, as they acknowledge the crowd's enthusiasm by waving their hands.

The sound-track to go with these pictures, silent yet thrilling, is provided by Rodzianko, the Speaker of the Duma, who recorded the scene: 'The crowd knelt to sing "God Save the Tsar", then stood up to shout, "Hurrah!" Because of their shouts, when the Tsar tried to say a few words, we could hear nothing but the crowd's "Hurrah", which drowned his voice.'

Incognito, the Speaker buttonholed a couple of workers in the crowd: 'What's happened to yo. strike, to your demands addressed to the Duma?'

'That was a quarrel in the family,' they replied, 'but now all Russia is involved. We have rallied to our Tsar and we shall follow him to victory over the Germans.'

They had defined in a single phrase the sacred unity of the nation which, as though by magic, gathered the court, the Duma and the parties round the throne. 'About 150,000 men had been on strike; there had been barricades in the poorer quarters ... But behold, war was declared; and all at once not a trace was left of the revolutionary movement. The workmen of St Petersburg returned to their factories' (Kerensky).

After the ceremony of the declaration of war, Grand Duke Nicholas, who was to be commander-in-chief, went up to the French Ambassador and embraced him, saying: 'God and Joan of Arc are with us! We shall win!'

Nicholas II had wished to assume command of the forces, but Prime Minister Goremykin and, especially, Sazonov, the Minister of

Foreign Affairs, dissuaded him: 'It's to be expected that we may be forced to retreat during the first few weeks. Your Majesty ought not to be exposed to the criticism such a retreat would be bound to give rise to.' The Tsar in protest referred to the example of Alexander I in 1812. Sazonov replied: 'If Your Majesty would graciously read the memoirs and correspondence of that period you would see how your august ancestor was criticized and blamed.'

The Tsar eventually bowed to their opinion. He appointed his uncle Nicholas to the supreme command, instead of General Sukhomlinov, the War Minister, who had wanted to be Generalissimo. The latter made his anger known.

No sooner had operations begun at the front than troubles began to stir in the rear. The Russians disliked all those Baltic barons and other nobles of German origin who thronged the court, such as Count Freedericks, the Minister of the Imperial Household, Baron Korff, the Grand Master of Ceremonies, General Grünewald, the Chief Equerry, Count Benckendorff, the High Marshal and all the other Meyendorffs, Budbergs, Kotzebues, and so on. Then there was the rumour that circulated in August 1914 that, if Rasputin had been present, he would have prevented the war. At the end of June he had been stabbed by one of his lady-friends, Khinya Guseva, who was denounced as a prostitute and locked up as a madwoman. Rasputin had gone off to convalesce in his native village, near Tobolsk in Siberia. Countess R. said to the French ambassador:

If the *starets* had been there, we should have had no war. I don't know what he would have done or advised but God would have inspired him, whereas the ministers have proved incapable of seeing or preventing anything ... Just look what determines the fate of empires. A harlot revenges herself on a dirty muzhik: the Tsar of Russia at once loses his head. And here's the whole world on fire!

It was said that the Tsaritsa sent Rasputin a telegram every day. What was certain was that the Tsarevich was in bad health. On the day war was declared he had not been able to leave his bed, and so the Tsar could not show his son to the people. 'Yes,' Nicholas commented a few days later, 'I am an unlucky Tsar.'

'But they have gone mad,' wrote the essayist and poetess Zinaida Hippius when she saw her compatriots drunk with enthusiasm and delirious as the soldiers marched off to war. The government itself was worried and perturbed. Prince Obolensky, commandant of St Petersburg, decided to put a stop to these patriotic demonstrations. Mobilization was not accompanied everywhere by the same excesses.[1] The traditional resistance to the call-up had reasserted itself. In some thirty districts there were clashes that left between 250 and 500 victims. But these were fewer than the authorities had feared. More or less everywhere, when the bells rang out and the Cossacks presented themselves for the enrolment, Holy Russia was there, to defend the sacred soil from Turk and Teuton.

A few weeks earlier, that same Zinaida Hippius had already noted another outbreak of madness, on the part of the strikers, at the beginning of the summer:

I was amazed by the disorders which occurred in Petersburg in the last days [before the war] . . . I understood absolutely nothing and I could feel that those who told me about it also understood nothing. And it was even clear that the rioting workers themselves understand nothing, even though they smash streetcars, stop traffic . . . And the intelligentsia [stood there] with its mouth hanging open – for them, all of this was like a snowfall in July.

The violence of these strikes and demonstrations, and the frenzy of this entry into war, seemed disquieting to Zinaida Hippius. And there were other surprising events.

At the beginning of 1914 the revolutionary and anti-war movement was a rising wave that, as it broke, seemed on the point of submerging everything. Yet now, within a few days, most of the revolutionaries rallied to the defence of the fatherland, headed by the Marxist Plekhanov and the anarchist Kropotkin. It was said that Lenin was

1. On the violence of the strikes in 1914 see the article by L. Haimson.

calling on the revolutionaries in every belligerent country to promote the defeat of their own government – but nobody in the Second International followed his lead. In the Fourth Duma the Bolsheviks alone refused to vote the war credits. But what weight had they? Apart from them, the *union sacrée* triumphed spontaneously. The situation had been turned upside-down.

Tsarism recovered its vigour and its legitimacy; 1914 was its year of glory.

If there should be defeat, and that began to seem a possibility at the end of the first year, if the economy's incapacity to supply the army's needs should be revealed or poverty should increase, then a revolutionary situation would reappear at once. And, this time, during a war and in face of an enemy much more dangerous than in 1905.

Lenin explained that, for revolution to succeed, two conditions have to be present: serious discontent among the masses and loss of confidence on the part of the rulers in their ability to react. It was the second feature that was the first to appear in 1915–17.

Ever since the violent events of 1905 a section of the intelligentsia had sensed the approach of 'a cataclysm', and some of them even hailed it. The new fact, from 1915 on, was that the ruling class itself, the government as well as the opposition in the Duma, listened with terror to the rumblings that drew nearer. The government and the population ceased to have any confidence in each other. As for the Duma, when the revolution came in February 1917, it did not know whether the demonstrations were for it or against it, and it could not see how to prevent the military catastrophe and the onrush of strikes and demonstrations in the capital. The Duma wanted to get rid of the 'incompetent' government but did not know how to do that. 'The Right declines to perceive the approaching catastrophe and prevents anything at all from being done about it. The Centre [meaning the Octobrists] keeps saying that the country is heading for disaster, but itself just marks time. The Left moves forward, but has no idea where it is going.'

Everyone agreed in making a scapegoat of Rasputin. Yet, when he had been murdered, at the end of 1916, it was seen that this event had changed nothing. Abroad, people sometimes said that Russia was a giant with feet of clay. But wasn't it, rather, a giant whose head was paralysed? War and revolution were to show.

The Russian Army had been regenerated since the defeats of 1904–5, thanks to the influence of General Bayov at the General Staff Academy and thanks also to the stimulus given by Grand Duke Nicholas, though he had been removed from the supreme command after 1908. Despite the succession of ten Chiefs of General Staff in ten years and, especially, the superiority of the Austrians and Germans in artillery, the Russian army showed up well in the first year of the war. It had the Grand Duke Nicholas at its head once more.

After a first great victory over the Germans at Gumbinnen and then another in Galicia, over the Austrians, it had suffered very heavy losses at Tannenberg. But the Russian Army had done its part in the east and thereby prevented Wilhelm II from obtaining a quick victory in the west. Without Gumbinnen, France would not have won at the Marne. But, just as the declaration of war had only seemed to sound the knell of the revolutionary movement, these first campaigns were illusory. In Russia as elsewhere it had been assumed that the war would be a short one. It was found that, tragically, the stocks of shells were sufficient only for a campaign of twelve weeks, and the factories' production could meet only a third of what was needed. In the rear, as the mobilization and transport of industry for war purposes took effect, shortages and a rise in the cost of living were not long in appearing.

At the same time, the personal success achieved by Grand Duke Nicholas began to worry the Tsaritsa. She was particularly unforgiving of the contempt he had shown for Rasputin. When the latter had indicated an intention to visit GHQ, the Grand Duke had replied: 'Yes, do come. I'll hang you.' As the campaign of 1915 began badly, the Tsaritsa was afraid that the responsibility for mistakes might, sooner or later, be laid at her husband's door. She wrote to the Tsar: 'Oh, I do not like N[icholas Nikolaevich] having anything to do with these big sittings [sic] that concern interior questions; he understands our country so little and imposes upon the ministers by his loud voice and gesticulations.' Besides, 'a man who turned simple traitor to a man of God's [*sic*] cannot be blest, nor his actions be good'.

The Tsar was, in fact, jealous of Grand Duke Nicholas. This giant of a man had such a presence! He was nearly two metres (6 feet 6 inches) tall, and he knew how to address the troops; whereas, when the Tsar needed to speak to them, a minister had to provide a script, hidden in his cap. The commander-in-chief was very popular. His reputation was doubtless overdone, and those close to him had noticed that, on the pretext that he presented such an easy target, he always stayed well away from the front. Nicholas II was much more courageous: in British documentaries he can be seen visiting wounded soldiers in the front line. He went there again and again, as though ready for sacrifice, but no bullet, not even a stray one, ever struck the Tsar.

Suddenly, in the summer of 1915, Nicholas decided to dismiss the commander-in-chief and assume the supreme command himself.

The Tsar had been urged to get rid of the Grand Duke by his War Minister, General Sukhomlinov, who put it to him that the commander-in-chief was too conciliatory in his dealings with the Duma representatives. He had allowed his generals to invite Guchkov and the members of the Duma's War Committee to visit the front, in order that they might help with the supply of munitions. 'This interference can become very dangerous,' Sukhomlinov explained. Nevertheless, the Tsar had agreed to the constitution of the committee in question, with a view to improving the supply situation, and had dismissed N. Maklakov, the Minister of the Interior who had hindered this co-operation.

The committee then took against Sukhomlinov, whom the Duma made the scapegoat for the army's shortages – an exaggerated accusation, no doubt, since the problem transcended the competence of the War Minister. A campaign against him was orchestrated by members of the Duma, which was due to meet on 19 July 1915. Despite Alexandra's support for Sukhomlinov, Nicholas yielded, before that date, to pressure from these men and from certain generals, and dismissed him on 11 June 1915. 'I had to sacrifice you,' he wrote in the letter he sent to the outgoing minister.

The news then came that, as a result of General Mackensen's offensive, the Germans had entered Warsaw. Pushing on, they were soon in Brest-Litovsk, then Kovno (Kaunas). There then began, amid indescribable disorder, the evacuation to the rear of whole populations. 'This immense migration organized by GHQ is bringing the country to disaster, to revolution, to perdition,' was the opinion of A. Krivoshein, the Minister of Agriculture.

News of the dismissal of Grand Duke Nicholas and his replacement in the supreme command by the Tsar himself reached the Council of Ministers at the same time as this discouraging information. From a minute kept by A. Yakhontov, the assistant to the head of the Chancellery, whose duty it was to record the council's proceedings, we know exactly how the ministers received the news and what their reactions were. This was the session of the council held on 6 August 1915.

They had discussed the evacuation from Poland of the Jewish element, regarded as potential spies, and who, if treated too badly, might take reprisals by stopping the loans to the government agreed to by the banks. They had gone on to the matter of soldiers' pensions, and then to the revolutionary agitation in the rear, which was aimed at making the workers suspect the government of treason.

General Polivanov, the new Minister of War, had taken little part in these discussions. 'He sat silently most of the time. It was clear that something was troubling him. The usual tic of his head and shoulder was particularly pronounced.' The chairman, Goremykin, eventually called on him to speak. 'With barely concealed excitement, in short abrupt sentences', Polivanov began his contribution:

The military situation has both worsened and grown more complicated ... Both at the front and in the rear of the armies we can expect an irreparable catastrophe at any moment. The army is no longer retreating, it is running away ... The appearance of a small German patrol evokes panic and results in the flight of whole regiments. So far, we are being saved from total disaster only by our artillery ... But its ranks are thinning, there are no shells left, and the guns are worn out. Headquarters have completely lost their head ...

But, no matter how terrible things are at the front, there is a far more dreadful event threatening Russia. I am deliberately going to reveal a military secret and break my word to keep quiet about it for a time. I feel compelled to warn the government that when I reported to His Majesty this morning he announced to me his intention to dismiss the Grand Duke and to assume, himself, the supreme command of our armies.

This revelation was 'another stunning blow in the midst of the military misfortunes and internal complications that were being suffered'.

'Knowing the sovereign's suspicious nature,' Polivanov went on, 'and his characteristic stubbornness when he has taken decisions of a personal order, I tried, with the utmost caution, to implore him at least to postpone the implementation of this decision.'

Prince N. B. Shcherbatov: I had heard rumours recently about intrigues at Tsarskoe Selo against the Grand Duke ... But I never thought that the blow would be struck at this moment, the most unsuitable moment for such a decision.

S. D. Sazonov (addressing the Prime Minister): How could you hide this danger from your colleagues in the Cabinet? After all, this issue affects matters on which the fate of Russia depends. If you had told us frankly, we would probably have found ways of counteracting the decision taken by the Emperor ...

I. L. Goremykin: I did not consider it possible to reveal that which the Emperor ordered me to keep secret ... I am a man of the old school. For me, an imperial order is law ... I must say to the Council of Ministers that all attempts to dissuade the Emperor will be, in any event, useless ... He has told me, more than once, that he will never forgive himself for not leading the army at the front during the Japanese war. His Majesty considers it the sacred duty of the Russian Tsar to be with his troops in time of danger, sharing both joy and sorrow ... Now, when the front is on the brink of catastrophe, His Majesty sees it as the sacred responsibility of the Tsar of Russia to be among his soldiers and with them either to conquer or to die ... Given these mystical feelings of his you will not be able, whatever arguments you use, to persuade the sovereign to hold back from the step he has decided on. I emphasize that no intrigues or influences of any sort have played a part in this decision of his ... We cannot do otherwise than bow to the will of our Tsar.

Eight ministers refused, and signed a petition directed against the Tsar's decision. Such a move was unprecedented, for the ministers did not emanate from the Duma but had almost all been chosen by Nicholas, by the Tsaritsa and by Goremykin.

'Don't give in and don't dismiss me,' N. Maklakov had said to the Tsar a few weeks before. 'If you dismiss me the liberals in the Duma will shout still louder.' They shouted louder than ever now that the voice of the Duma was finding expression within the government, which was looking ahead and foreseeing a catastrophe.

Nicholas began by ignoring the petition from his ministers. Then, on 2 September, he prorogued the Duma. On 4 September the Duma's War Committee, headed by Shingarev and Shulgin (a Cadet and a member of the Union of the Russian People) sent him an anguished appeal to change his mind.

Sire,

The Russian armies, and all Russia along with them, are now passing through severe trials, and this is what has led us to appeal to you, Sire, to accept, if you will, the present memorandum. We have set forth here, in abridged form, everything we have learned about the war and the way it is being conducted that has resulted in the difficult situation we face today, and have also stated what we consider can remedy the grave misfortune that has come upon our country ...

We have learned that our valiant army, after losing more than four million men, killed, wounded or taken prisoner by the enemy, is not only retreating but will perhaps have to withdraw still further. We have also learned the reasons for this retreat, which is such a cause of sorrow to us. We have learned that our army, in its fight with the enemy, does not possess armaments of equal effectiveness, and that, while the enemy ceaselessly showers upon us a hail of steel and lead, we are unable to send him in reply more than a much smaller quantity of bullets and shells.

We have further learned that, while the enemy possesses plenty of artillery, both light and heavy, we are almost completely lacking in these weapons, and that, as regards our light cannon, these have been used so much that they will soon become unserviceable.

Again, we have learned that, whereas the enemy is increasing daily the number of his machine-guns, to the point where, according to the information of the War Ministry, he has a formidable 55,000 of them, we have barely enough to replace the ones we have lost or have had to scrap.

We have learned, furthermore, that, although the enemy has plenty of rifles, and each of his soldiers is equipped with one, hundreds of thousands of our men are without weapons, and have to wait until they can pick up the rifles dropped by their fallen comrades.

We have learned, too, that, while many things in this war have exceeded the bounds of human understanding and could not have been foreseen, it would nevertheless have been possible to avoid many other things if there had not been criminal negligence on the part of certain military commanders.

Then, we have learned that the front sent back, so early as September of last year, a warning that the supply of shells would not be sufficient and a request that steps be taken in good time to see to this matter, but that no action was taken. Only when the danger drew near and became inescapable was a move made to deal with it. But we know that several months yet will be needed, not even to come level with the enemy but just to draw near to him in armaments.

We have learned other things as well. We have learned about the conditions under which our troops retreated from Galicia. We have learned that nowhere did our forces find strong positions prepared in advance. We have learned that, after wearisome marches, our men themselves had hastily to dig pitiful trenches, as well as they could, and wait in them for the enemy to catch up and hurl the hurricane of his heavy artillery bombardment on to our exhausted, worn-out soldiers, for whom the ditches they had just scratched out offered no protection.

We have learned that even the most important places, the great cities of our native land, were either insufficiently fortified or not fortified at all ...

We have learned that appointments to important military posts – as, for example, the command of divisions and corps – have been made in accordance with seniority of rank, following a special list which records generals' seniority, with exceptions made only for officers with powerful protectors.

It is thus neither bravery, nor talent, nor competence, nor military worth, proved by facts, that is decisive in the promotion of candidates, but other considerations altogether. Under such conditions, really able persons, true leaders of men who could lead our troops to victory, attain only rarely the highest positions in the army. Generally speaking, those positions are allotted to officers with seniority, even if they are less gifted. Yet three-quarters, perhaps, of success in the art of war depends on judicious selection of leaders. For this reason the present method of appointment is disastrous for our cause.

All this we have learned, Sire, but we have also learned something still more deplorable, namely, that all these woes and all these muddles have begun to affect the morale of the army and of the people. Perceiving as they do the negligence, carelessness and lack of organization that prevail everywhere, the troops have started to lose confidence in their leaders ...

A similar mood of distrust, discontent and exasperation is manifest, and in even stronger forms, among the people themselves. The people know that shells and bullets are available in very limited amounts, they know that this is somebody's fault, but they see that public initiative is pressing vigorously forward to correct the former mistakes, to make up for lost time and to focus all the forces that can help to equip the army properly. It is therefore not the errors of the past that are worrying the people today.

What the people do not understand is why trenches are not prepared in sufficient numbers, and only when the enemy is already near, and why this is done in such disorderly fashion. They do not understand why, after tens of thousands of men have been assembled, they are left idle for days on end, or simply sent back to where they came from.

The people are ready to work for the defence of our country. They will dig with a will every foot of Russian soil in order to protect our native land with an impassable line of fortifications, just as our French allies have done ...

Your Imperial Majesty! We make bold to say to you that we appreciate the inevitable disjunction between the authority at the head of the army and that which governs the country. Yet we are deeply convinced that it is impossible to conduct national defence successfully without a supreme authority that unites everything. The indisputable power of the Tsar is the only authority that can establish accord between GHQ and the government.

The Tsar can order the civil and military leaders to work out in advance a plan of action for a long period ahead, taking account of all the complexity and diversity of the consequences that must follow from decisions already adopted, so as to put an end to these disordered moves that look only a few days ahead and lack any underlying general idea.

The Tsar can broaden the scope of calculations and considerations. The Tsar can insist that the aim taken is not to imitate the enemy in timid fashion but to bring to bear all the strength of an immense and powerful country that wishes to win at any cost, so as to surpass the enemy in both equipment and foresight.

The Tsar alone can prescribe that important positions be given to persons who have proved their valour in battle and not to persons who are often incapable of carrying out successfully the hard tasks of war. The Tsar can bring together all the forces of great Russia in order to create impenetrable lines of defence that will safeguard our country until the supreme moment comes when Providence will choose to grant us final victory over an enemy who will have been worn down by our firmness.

Sire, we believe in that victory.

This document was signed by 'loyal subjects of His Imperial Majesty', A. Shingarev, chairman of the War Committee, and this committee's seven members, and dated 4 September 1915.

The Tsar failed to reply to this appeal and his ministers, in despair, measured the danger that was beginning to become clear as it hovered over them. But they were too short-sighted to see any further than the Duma.

'The army and the population rely no longer on us but on the War Industries Committee and the Duma,' said P. Kharitonov.

Sazonov said: 'The government is hanging in mid-air and satisfying no one.'

Prince Shcherbatov exclaimed: 'How do you expect to fight the growing revolutionary movement when I am refused the co-operation of the troops because they are supposedly unreliable?'

The evacuation of the Jews from Poland, about which the Duma had argued, was completed more successfully than it had been begun. Goremykin convoked a session of the Council of Ministers at GHQ. 'You foretold the worst ... You said that there would be a revolution if I prorogued the Duma. Well, there hasn't been.'

And the Tsar said to them: 'An order has been given me from on high ... I well remember how, when I stood before the great icon of Our Lord in our chapel at Tsarskoe Selo, an inner voice called on me to take the supreme command and to inform the Grand Duke, independently of all that had been said to me by our Friend [that is, Rasputin].'

On the eve of this meeting of the council he had received a letter from Alexandra (15 September 1915): 'My very own precious Darling ... Remember to keep the Image in your hand again and several times to comb your hair with His comb before the sitting [*sic*] of the ministers. Oh, how I shall think of you and pray for you more than ever then.'

Nicholas stood up to his ministers and on the 16th dismissed the lot of them.

David R. Jones has correctly shown that Nicholas II's decision to assume the supreme command, replacing Grand Duke Nicholas, was not due merely to his envy of the Grand Duke or to any pressure from Sukhomlinov, Alexandra or Rasputin. It resulted from more fundamental considerations. The incompetent Grand Duke had to be replaced without embarrassing the dynasty, and only if the Emperor himself took over the supreme command could that difficulty be overcome. This action could put an end to the calumnies that were circulating about his alleged affinities (through Alexandra) with the Germans. Finally he considered that, as autocrat, he should merge with his people, sharing day-by-day the same dangers. His ministers told him that he was taking a risk; that, if he were defeated, he would drag the dynasty down with him. For his part, he might well think that, on the other hand, if he was victorious, the autocracy would have won the game.

'Be more autocratic, my very own sweetheart, and show your mind.' Messages of this sort reached Nicholas almost every day when he was at the front. 'You have fought this great fight for your country and throne – alone and with bravery and decision. Never have they seen such firmness in you before.' Thus wrote Alexandra after the Tsar had prorogued the Duma. She added: 'Do not fear for what remains behind ... Lovey, I am here, don't laugh at silly old wify, but she has "trousers" on unseen, and I can get the old man [Goremykin] to come and keep him up to be energetic.'

In fact, while Nicholas commanded the army at the front, Alexandra wielded power in the rear. Her correspondence testifies that she was into everything: she supervised appointments, consulted and issued orders. The 400 letters she sent, on a daily basis, to the Tsar are reports on the affairs of the court and the government: long letters in which were mingled details about the children, gossip and advice on strategy. She knew what she wanted. Between 11 September 1915 and 15 March 1916 she dispatched fifteen letters in which she urged Nicholas to make N. Khvostov Minister of the Interior. Nor did she fail, whenever occasion arose, to impart her view, which was often that of Rasputin, on the Duma, on Brusilov's offensive, or whatever.

In this way she, along with Rasputin, obtained the appointment of several ministers. For example, she explains to Nicholas: 'Darling, remember that it is not a matter of the man Protopopov or XYZ, but it is now a question of monarchy and your prestige, which must not be shattered by the Duma.' 'Be Peter the Great, John the Terrible, Emperor Paul ...' 'With a clear conscience, in face of all Russia I would have sent Lvov to Siberia ... taken Samarin's rank away ... and sent Milyukov, Guchkov and Polivanov also to Siberia.'

In a letter to Alexandra dated 13 September 1915, Nicholas shares with her the anxieties he feels:

The military situation ... looks threatening in the direction of Dvinsk and Vilna, grave in the centre ... and good in the south (Gen. Ivanov) ... The gravity lies in the terribly weak condition of our regiments, which consist of less than a quarter of their normal strength; it is impossible to reinforce them in less than a month's time, as the recruits will not be ready, and, moreover, there are very few rifles ... I beg you, my love, do not communicate these details to anyone; I have written them only for you.

It is all there in this short letter – the tragic balance sheet of the summer of 1915 (nearly 2 million victims), the shortage of men and, even more, the lack of rifles. One feature of the scene is missing, though; there are no longer any shells for the guns. True, all the belligerents experienced something like this, since none had expected that the war would last so long. Nowhere, though, was the shortage of munitions so dramatic as in Russia. The command had to offset this shortage by ordering crazy bayonet charges aimed at deceiving the enemy as to the real situation in the army. Though it managed to effect a certain degree of deception, it was forced to abandon all Poland.

To this letter Alexandra replied: 'I tell nobody what you write to me, except Him who protects you wherever you are.' 'Him' meant Rasputin, who knew all the secrets.

The Minister of the Interior at that time, A. N. Khvostov, told the commission set up by the Provisional Government, after the February Revolution, that the banker Rubinstein had asked Rasputin to find out, when he went to Tsarskoe Selo, whether the Russian Army was going to launch an offensive. He needed to know that because he was thinking of buying some forests in Minsk province. If there was to be an offensive he would go ahead and buy them, as they would be very profitable, but otherwise not. When Rasputin returned from Tsarskoe Selo, Rubinstein got him drunk and he told how the Tsar had been very depressed because of the army's shortage of boots and rifles, so that there could be no immediate offensive. When Rasputin asked when the Tsar thought that an offensive would become possible, the reply was: 'Not for another two months: we shan't have enough rifles before then.'

Whether true or not, this statement, together with the letters exchanged between the imperial couple, shows how information circulated and how decisions were taken. At the time, the term 'camarilla' was applied to the little group of persons in the know who, with Alexandra, Rasputin, Vyrubova and the ministers who owed their appointment to Rasputin, were pursuing a policy that, it was thought, was leading the country to ruin. The central authority was increasingly isolated, from the country, from the Duma, even from the ministers.

The defeats of 1915, the replacement of Grand Duke Nicholas at the head of the army by the Tsar himself, the crisis of supplies in the rear, the shortage of munitions at the front, the disquiet inspired by so much negligence, all this stirred up revolt among the 'active classes'. The business bourgeoisie, the members of the zemstvos and the army command, sensing the onset of danger – danger of revolution in the midst of war, above all – worked up an offensive directed against the person of Nicholas II, 'the incapable Tsar'. It was necessary to replace him in order to save Russia and Tsardom.

This idea arose after the appearance in *Russkie Vedomosti* in the autumn of 1915 of an article by V. Maklakov entitled: 'A tragic situation: the mad chauffeur'. This famous parable circulated throughout Russia.

Imagine that you are in a car that is racing headlong down a steep, narrow road: one wrong turn and you will be gone for ever ... Suddenly you realize that the chauffeur cannot drive. Either he does not know how to handle the car on steep hills, or he is tired and no longer understands what needs to be done. In any event, he is carrying you and himself to destruction. Fortunately, there are people in the car who know how to drive. Consequently, it is necessary that they take over from the chauffeur as soon as possible. But doing this is highly dangerous when the car is travelling at such a speed. Moreover, either through blindness or from professional self-respect, the chauffeur stubbornly refuses to give up the wheel and will not let anyone else take it. What is to be done? One move of his hand could hurl the vehicle into the abyss. You know that and so does he. And he laughs at your anguish and your helplessness. 'You won't dare touch me.' And he is right – you won't dare. More than that. Not only will you not interfere with him, you will help him with your advice. And you will be right. That's what must be done. But how will you feel if you realize that, even with your help, the chauffeur cannot drive, and if your mother, seeing the danger, begs you to help him and accuses you of cowardly indifference?

The chauffeur was Nicholas II; the mother, Holy Russia; the competent persons, those 'associations', formed privately but concerned with public matters, that were gradually taking charge of the country's interests.

Faced with the 'negligence' of the persons in power, as people put it at the time, associations had been formed and had obtained from the Tsar or his ministers permission to help the government save Russia. Naturally, they had to operate cautiously, lest they offend a bureaucracy jealous of its prerogatives. The Red Cross Committee had set the example. A modest affair to start with, it had gradually taken over the administration of the country's hospitals. The zemstvos had also come together in a Union of Zemstvos which was presided over by Prince Lvov, in order to co-ordinate the reception of refugees, the placing of prisoners of war, and so on. Then came the War Industries Committee, guided and chaired by Guchkov, the purpose of which was to rationalize defence production. At the same time, as a result of the shortage of goods in the rear, consumers had created a network of co-operatives; by 1917 there were 35,000 of them, with more than 10 million members.

These initiatives bore witness to the vitality of Russian society, but the government eyed them askance. Little by little the administration saw itself losing its functions, and was unable to check the process. Without being aware of it, the Russians were starting to govern themselves: the army on one side, the producers, or the consumers, on the other. Revolution was not yet in the people's minds but it was beginning to take shape in the facts of Russian life.

Like the associations and the zemstvos, the legal opposition was still timid. In any case, the same men and the same ideas were playing the same roles – in the Duma, especially, where the truce concluded in 1914 was broken. Inspired by the Octobrists, a majority of the deputies formed a 'Progressive Bloc', which was joined by some members of the State Council and ministers. The aims of the bloc were moderate, in that it did not even dare to ask for a ministry responsible to the Duma, but merely a government that would have the Duma's 'confidence'. It did, of course, call for changes in methods of government, an end to the prosecution of persons who had committed no crime, the return of administrative exiles, no more religious persecution, abolition of the anti-Jewish measures, restoration of the Ukrainian press, of the trade unions, of the friendly societies, and so on. Prime Minister Goremykin, ever vigilant where the prerogatives of the autocracy were concerned, disapproved of the constitution of this bloc, which had been formed 'illegally'. The bloc then launched a campaign against the government, in which the Cadets joined. The Prime Minister considered all this noise inopportune and illegitimate; the situation, as he saw it, was not so grave as all that.

'Either the government is concealing the truth from us and deceiving us, or else it is blind, and this is the mark of its incompetence,' said Milyukov in the Duma, to thunderous applause. Goremykin closed that session of the Duma.

The anger of Guchkov, Milyukov, Maklakov and Rodzianko, the Speaker of the Duma, was all the stronger because what they were trying to do was to prevent the revival of the illegal opposition that all feared was likely, so difficult was the situation becoming, in the rear as at the front. Shortages, the fall in purchasing power and repression were arousing a discontent of which a sign was the rise of a strike movement that had resumed on an extraordinarily large scale.

The synchronizing of demonstrations showed that there must be a central organization, close relations having been restored between the underground movements in Russia – Bolsheviks, Mensheviks, Socialist Revolutionaries, anarchists – and their leaderships abroad. 'The military defeats contribute to the collapse of Tsardom,' Lenin wrote from Switzerland to Shlyapnikov, who was in Russia. 'They facilitate the union of revolutionary workers.'

He was at this time writing *Imperialism, the Highest Stage of Capitalism*, in which he argued that, contrary to what socialists had supposed before 1914, the revolution would break out not in the country where capitalism was strongest but in one that was not greatly developed economically and could not sustain its war effort. This reversed the terms of Marxist dogma, and signified that the explosion was more probable in Russia than elsewhere. At the end

of 1916 the legal opposition certainly did not think that the time for
this had arrived. But, since it considered the Tsar to be incapable, it
estimated that the time had come to act so as to prevent the worst.

As if they wished to provoke the Duma, Nicholas and Alexandra
chose, to succeed Goremykin, one of Rasputin's 'creatures', a provincial governor named Stürmer, an old Okhrana hand, who called
himself a reactionary but brought discredit on the extreme Right.
That tendency fulminated, through its leader Purishkevich, against
the 'camarilla', whom it accused of being Germanophil and favourable to a separate peace. Everyone denounced Rasputin.

In the course of a few years Rasputin had enlarged the bounds of
his empire. According to Father Shavelsky, when court chaplain, the
starets wielded over the Tsaritsa an unrivalled moral ascendancy.
This was not equally true, perhaps, of Nicholas, but he was gradually
being drawn into Alexandra's circle of mystics and getting more and
more dominated by them.

Rasputin's antechamber had become a sort of waiting-room where
place-seekers of every grade jostled each other – this one to be made
Metropolitan or general, that one to be made a minister, even. There,
too, were to be found unfortunates of all kinds who came to beg his
help. The élites of the country, and even more those of the court,
grew ever angrier as they realized that the 'holy man' was gradually
filling the state apparatus with his 'creatures'. Even if Sazonov,
Kokovtsov and some other ministers, like Stolypin before them,
owed nothing to Rasputin and ostentatiously ignored him, his supporters none the less thronged the monarch's rooms. Among such
were the teacher of divinity to the imperial children, Archpriest
Vasilev; General Voyeikov, who commanded the palace guard;
Metropolitan Pitirim; the head of the court Chancellery, Taneyev;
and the latter's daughter, Madame Vyrubova, the Tsaritsa's companion. Soon, too, there were a couple of ministers from the same
set. In 1916 Prime Minister B. Stürmer and Minister of the Interior
Protopopov figured among the adepts of Rasputin's spiritualistic
seances. The scandal caused by his debauchery gave him no anxiety,
for his standing with Alexandra was unshakeable owing to the help
he was able to give to the Tsarevich. At the front, to which he

accompanied his father in 1915, Alexei's nose started to bleed; he had to be brought back without delay to Tsarskoe Selo, and Rasputin once more showed himself more effective as a healer than the official doctors.

His rise had caused envy and hatred. At court and in the city he had made many enemies, especially among those persons who felt excluded from the influential circle around him. The Okhrana even put him under surveillance, and the presence of one of Rasputin's 'creatures' at the Ministry of the Interior made no difference to that arrangement. The police agents used spy-holes to observe what went on in at least some of the places where he amused himself.

The country was in the grip of the 'Rasputinshchina'.

A conversation between the Tsar and the chaplain-general of the army and navy in autumn 1916 shows how discredit was affecting the regime, and that Nicholas was aware of this. Father Shavelsky had succeeded in obtaining an audience with the Tsar in order to talk to him about various matters, including Rasputin's misdeeds. Actually, he just repeated what was being said everywhere, rather than saying anything either precise or fresh. Then he continued:

'Do you know, Your Majesty, what is going on in the country, in the army, in the Duma? Do you deign to read the proceedings of the Duma?'

'Yes, I read them,' the Tsar replied.

'Have you read the speeches of Milyukov and Shulgin?'

'Yes,' was the answer.

'In that case, Your Majesty, you know what is happening in the Duma. Where attitude to the government is concerned there are now neither left-wing nor right-wing parties; all the Rights and Lefts have united into a single party discontented with the government, hostile to it ... Do you know that in the Duma they have openly called the chairman of the Council of Ministers a traitor and a thief?'

'How vile!' the Tsar exclaimed indignantly.

'If it is true, is it not justified?'

'But how can such nonsense be justified? ... I have known Stürmer a long time. I knew him when he was still Governor of Yaroslavl,' said the Tsar.

Shavelsky went on to mention that the closest colleagues of Protopopov claimed that he was mad. The Tsar, 'somewhat agitated',
replied:

I have heard that. Since when has Protopopov been mad? Since I appointed
him minister? Why, it wasn't I who elected him to the Duma, but one of
the provinces. The representatives of the nobility in Simbirsk province
elected him. The Duma elected him as their deputy chairman and then
chose him to head the commission they sent to London. Was he not mad
at that time? But as soon as I chose him to be one of my ministers everybody
started shouting that he was out of his mind.

The Tsar's obstinacy, his devotion to the Tsaritsa and their shared
faith in Rasputin eventually created a climate of unreality, as government and court circles came to attribute to Rasputin an activity and
responsibility in political and military spheres that he did not have
in fact – even though, as was well known, he considered that the
effect of the war would be to bring down Tsardom and that an
alliance with Germany would be altogether more logical than one
with the French Republic. P. Durnovo had said the same thing a few
years earlier.

In reality, Nicholas II remained loyal to his allies and, with
or without Rasputin, he marched with his eyes closed towards
catastrophe.

Despite the momentary success of Brusilov's offensive, people felt
in 1916 that the entire economic system was cracking and that the
army would soon be without supplies. Political criticism poured in
from all sides with greater violence and acrimony than in any other
of the belligerent countries. General Denikin tells us that a socialist
deputy to the Duma who was invited to visit the army was so
taken aback by the freedom with which 'the officers talked among
themselves about the baseness of the government and the moral filth
that prevailed at court, that he suspected they were trying to lure
him into a trap'.

Writing of the 'mess' that Russia was in, French Ambassador
Cambon reported that at the beginning of 1917 Petrograd made

General de Castelnau think of a madhouse. For the Allied conference on war aims, at which the Tsar was expected to recognize the legitimacy of France's claim to Alsace-Lorraine in exchange for a recognition of Russia's 'special interests' in the Straits, the French had decided to send Castelnau as their representative because, being an aristocrat, he could use the opportunity to talk freely to the Tsar about the need to seek support in the Duma. Samuel Hoare, the British delegate to this conference, published the memorandum that was handed to it by P. Struve, explaining that public opinion was convinced (wrongly) that Alexandra and her 'camarilla' were pro-German, and arguing that only a government formed by agreement with the Duma would shift that burden and end that suspicion. 'Postponing' the session of the Duma that was due to begin on 14 February, this memorandum stated, would inevitably lead to serious incidents. A group of deputies had arranged for this memorandum to be presented to the Allied representatives in the hope that, in the common interest, they would intervene with the Tsar.

But General de Castelnau realized very quickly that if he were to talk like that to the Tsar it would be the surest way to doom his war-aims mission and that, unless the Tsar were to address a question to him, he ought not to open his mouth.

Pressed by the Dowager Empress and the Tsar's sister Olga, Grand Duke Nicholas decided to send the Tsar, on 1 November 1916, this final appeal, which was also a warning:

Were you to succeed in eliminating this continual interference by sinister forces, Russia would feel she was reborn and you would regain the lost confidence of the great majority of your subjects ... When the moment comes – and it is near – you will be able, from your throne, to proclaim the responsibility of ministers both to you and to the legislature. That will happen quite simply, by itself, without any external constraint, otherwise than was the case with the memorable manifesto of 17 October 1905. I have hesitated for a long time to tell you this truth, and I do it now because your revered mother and your sisters have convinced me that I should. You stand on the eve of a new period of troubles – I will even say, of fresh attempts at assassination. Believe me, if I insist so strongly on the need for you to free yourself from the shackles that have been formed around you, I am not doing it for personal reasons ... but solely in the hope of saving you, your throne and our dear country from irreparable consequences.

When he murdered Rasputin during the night of 16–17 December 1916, Prince Yusupov meant to perform Act One in the regeneration of Russia. Freed from the *starets*, the Tsar would at last listen to the voice of the nation, that is, of the Duma, summon up his energies, win the war and restore the country. This was also the inspiration of his friends, Grand Dukes and Duma deputies, the latter including men of the extreme Right, like V. Purishkevich, as well as of the liberal wing, like V. Maklakov, A. Guchkov and Speaker Rodzianko, who were all exasperated with the growing influence of Rasputin and his clique.

In fact, instead of a 'regeneration', Russia carried out a revolution, the most total revolution ever known, and for some the murder of Rasputin was the first sign of what was coming. This was why, in 1919, Prince Yusupov reversed the significance he had assigned to his action. 'Rasputin', he said, 'embodied Bolshevism on the march.' In this manner the Prince presented himself as being the first among those who tried to prevent the triumph of Bolshevism.

Paradoxically, this interpretation was taken up in 1978 by the dissident historian Andrei Amalrik, who noted that there was a certain degree of identity between Rasputin's aims and Lenin's political programme: the land to the peasants, peace with Germany, equality of rights for the non-Russian nationalities. In fact, however, the analogy is deceptive. The measures that Lenin took in October 1917 in order to save the revolution and keep himself in power had been advocated by Rasputin a year earlier for the opposite reasons, namely, to prevent revolution and save Tsardom.

In the arguments he set forth to explain what he did, Prince Yusupov, a pleasant youth, did not take that line. He revealed, quite simply, the point of view of his family, namely, that the Romanovs preferred the company of a muzhik to that of their own cousins, that the *starets* had the entrée to Tsarskoe Selo, and Yusupov was furious that this influence Rasputin had acquired enabled him to choose governors and ministers. He wished to free the Tsar from this pernicious influence, decapitate the 'German party' (Alexandra, Protopopov, and the others) and strike down a man whose intrigues and immorality were 'dishonouring' the dynasty.

Actually, Alexandra herself wanted to restore good relations between her husband and his cousins. Protopopov and Stürmer, like Rasputin, thought that continuing the war risked jeopardizing the dynasty, but there is no proof that the Tsaritsa played any special role in the negotiations which Protopopov allegedly entered into at Copenhagen. All that we know is that the Danish court was the obvious place for talking about family matters or questions of international politics. But it seems clear that the Tsar himself knew nothing of any such moves. Following the best tradition of plots and assassinations at the Tsar's court, Prince Yusupov has told us how the murder of Rasputin was prepared, and the details of the crime. He says that, even though he had been warned, he himself succumbed to Rasputin's hypnotic spell. He invited him to an evening party at his house, where it was intended that armed men would deal with him. At the critical moment, when the *starets*, already poisoned, had been shot several times, Yusupov was astounded to see him get up and threaten his attackers, who had all thought him dead. Rasputin was certainly a force of nature and a man of formidable temperament.

The Tsaritsa collapsed when she heard of Rasputin's death, and Gilliard, her children's tutor, tells us: 'Her agonized features betrayed, in spite of all her efforts, how terribly she was suffering ... He who alone could save her son had been slain ... The period of waiting began – that dreadful waiting for the disaster that there was no escaping.'

As for Nicholas, who was not at Tsarskoe Selo but at GHQ, a witness tells us that when he was told of the death of the 'holy man' he walked away whistling cheerfully. He had begun to feel Rasputin's ascendancy as a burden. A few weeks before, he had written to Alexandra about changes in the composition of the Council of Ministers: 'Only I beg you, do not drag our Friend into this. The responsibility is with me, and therefore I wish to be free in my choice'

(10 November 1916). He had already, on 9 September, warned her that 'our Friend's opinions of people are sometimes very strange, as you know yourself – therefore one must be careful, especially with appointments to high offices'.

All the same, Nicholas ordered that the murderers of Rasputin be subjected to severe punishment, in the form of exile. After the victim's body had been found by the police, under the ice in the Neva, Prince Yusupov was allowed to go abroad, but Grand Duke Dmitri Pavlovich was sent into exile at the front in Persia.

'Things can't go on like this; it makes me think of the days of the Borgias,' Grand Duke Nicholas Nikolaevich was heard to say. Actually, as in Milan, a conspiracy was being organized within the family, a conspiracy involving Grand Dukes Gabriel Konstantinovich, Cyril, Boris and Andrei and the Dowager Empress.

The main idea was to get rid of Alexandra, who was seen as being responsible for all Russia's woes, and was openly accused of sympathy with Germany. Influenced by Rasputin, and herself persuaded of the dangers to be feared from a protracted war, the Tsaritsa did indeed think that it would be desirable to end the conflict. But there is no proof that she did anything to bring about that result. On the contrary, when, in the midst of the war, her brother 'Ernie' came to Russia secretly on behalf of the Kaiser, she refused even to discuss the idea with him.

She was, of course, German by birth, but she was English by upbringing and inclination, and Russian even more than that, because her husband and her children were Russian. Also, she detested Wilhelm II. At the very beginning of the war she expressed her feelings plainly: 'I am ashamed of being German,' she said when she was told of the atrocities committed by the Kaiser's soldiers in occupied Belgium. She threw herself, body and soul, into the management of the military hospitals, proving genuine devotion. This woman who had been so indolent before the war showed herself most energetic now. She wanted to be Matushka, the 'Little Mother' of Russia's soldiers. For the imperial family and for Russians generally, however, she remained Nemka, 'the German woman'.

What had given some substance to the charge of pro-Germanism was the arrest and execution as a German spy, on the initiative of General Yanushkevich, chief of staff to Grand Duke Nicholas, of one S. Myasoedov, who had been a colleague of Sukhomlinov's and whom Alexandra had defended. However, it appears that Myasoedov was innocent, just as Beilis had been innocent; but whereas the liberals had made it a point of honour to defend Beilis, they had turned themselves into prosecutors where Myasoedov was concerned. They were only too happy to be able to say that treason reigned, from the court to the armies, and that this could be blamed for the defeats suffered by the Tenth Army during the retreat in 1915. The affair made all the bigger stir because Guchkov had fought a duel some years previously for having said that Myasoedov was a traitor.

The dominance of Alexandra had become unbearable for the imperial family, and the Dowager Empress swore that she would not set foot in Tsarskoe Selo so long as her daughter-in-law was there. The conspiracy aimed, accordingly, at exiling Alexandra to the Crimea and acting together with the army and the organizations led by Guchkov in order to establish a regency under either Grand Duke Nicholas or the Tsar's brother Michael.

The Tsar seemed deafer than ever to all suggestions that he form a government enjoying the confidence of the Duma. This wounding proposal, which was aimed at him from all sides, became unendurable, and he saw it as a conspiracy, whereas for everyone else it was the only path to salvation. Discontent was rumbling in Petrograd, stirred up by the philippics in the Duma, and the liberals feared the worst, namely, the revolution that they thought a popular government could alone prevent.

In order to show just how he felt, Nicholas resolved to remove the members of the Progressive Bloc from the State Council and replace them with men of the Right, who would thus recover their majority. He made no reply to the wishes of the assembly of the nobility in Novgorod for 'the departure of the dark forces' and 'a government of confidence'. To Basilevsky, marshal of the nobility of Moscow, who voiced the same request, he said: 'Send the Moscow nobility my sincere appreciation of their prayers and expressions of feeling. Also tell them that no one is more grieved than I with the internal situation at such a moment when we must still fight with the enemy and when a close unity is necessary to deal him a final blow.' When he received Rodzianko, who warned him of the revolution he felt was approaching, Nicholas replied: 'My information indicates a completely different picture ... If the Duma allows itself to make such sharp attacks as the last time, then it will be dissolved.' On 10 February 1917, in the presence of Alexandra and Grand Duke Michael, Grand Duke Alexander Mikhailovich told the imperial couple that he could really see no solution other than a ministry acceptable to the Duma.

Alexandra sneered. 'All this talk of yours is ridiculous. Nicky is an autocrat. How could he share his divine rights with a parliament?'

Grand Duke Alexander replied, with emotion: 'I realize that you are willing to perish and that your husband feels the same way, but what about us? Must we all suffer for your blind stubbornness? No, Alex, you have no right to drag your relatives with you over a precipice!'

Thus, the Tsar's relatives had abandoned him even before the revolution broke out. At the same time they strengthened their ties with Guchkov and the military.

As chairman of the War Industries Committee, centred in Moscow, Guchkov was in constant contact with the military. Unlike Milyukov, who intended to wage the battle in the Duma, following the parliamentary road, Guchkov, backed by the military and the 'provincials', took the view that it was necessary to find support in the civic organizations, the zemstvos and committees, and also to seek additional backing to the left of the Cadets, among the more moderate socialists, if need be. These tactical procedures, though different, had the same objective – to carry out a palace revolution so as to prevent a real one happening, especially in the midst of the war.

A refusal by Prime Minister Stürmer to place a contract for rifles in Britain gave the signal for attack. Guchkov circulated a letter in which he wrote: 'When you think that this government is headed by a man, Stürmer, who, if he is not a traitor already, is getting ready to become one, you will understand the anxiety we feel concerning the future of the motherland.' The original of his letter had been sent to General Alexeyev. When Nicholas learned of this, he warned the chief of staff that this correspondence with Guchkov must cease. Nicholas had already been angered when Alexeyev forbade Rasputin to visit the troops at the front. This time Nicholas showed mistrust; Alexeyev fell into a nervous depression and had to go to the Crimea for a rest. His temporary replacement was General V. Gurko (November 1916).

At the New Year, Grand Duke Nicholas, who had been relegated to command of the armies operating on the Caucasian front, received, through the Mayor of Tiflis (Tbilisi), a proposal that he take over from the Tsar when the conspiracy should come to a head. The Grand Duke declined, considering that 'in the midst of war the country would not understand', but he did not think the idea was bad in itself, and he said nothing to the Tsar about it. Generals Brusilov and Ruzsky, the former of whom was covered with glory owing to his victories on the Galician front in the summer of 1916, gave their approval to Guchkov's plan. 'If we have to choose between the Emperor and Russia, we choose Russia.' Alexei would succeed his father, with his uncle Michael as regent.

We are told by Maurice Paléologue, the French Ambassador, who saw Nicholas at the beginning of 1917, that the Tsar's morale had gone, that he had lost faith in his mission. Like General Alexeyev he fell ill, which was a way of abdicating. Rasputin's death had not affected him personally, but he felt that he was becoming more and more isolated. His mother and his uncles were against him, and the extreme Right had applauded the murder of Rasputin.

In mid-February the Tsar left GHQ for a brief visit to his wife and his children, who were sick. Then, unexpectedly, on 22 February, he returned to Mogilev. Despite the separation from Alexandra, Nicholas enjoyed being with the army. He did not interfere in the decisions of the high command – which meant those taken by Alexeyev, on his return from the Crimea – but he visited the front, inspected the soldiers and comforted the wounded. This ritual he performed with conviction, and he took the heir to the throne with him, proud to be teaching the boy his future role, whereas Nicholas's own father had neglected that aspect of his education. However, his presence with the troops had another connotation, one that was much more dramatic.

We learn from the memoirs of General Dubensky, historian of the Tsar, that General Spiridovich, a former head of security, came hurriedly to the Crimea, where some of the imperial family were staying, to warn his former superior, General Voyeikov, about rumours of a plot that had reached his ears at Livadia. This was a plot to murder the Tsaritsa and Anna Vyrubova. General Voyeikov paid no attention to the story. The Tsar, however, had an intuition that something was going on, in the army at least, when his brother Michael told him of the ill feeling at GHQ caused by his protracted absence. The command considered that when he left for Tsarskoe Selo, Nicholas fell again under the influence of the 'camarilla'. The Tsar had been aware of pressure upon him by the Allies during the conference in Petrograd in January. He knew that the British Ambassador, Sir George Buchanan, was in constant contact with Guchkov, Milyukov and the Grand Dukes. He certainly did not realize that their common target was Alexandra, or he would not have left her side. Consequently he acted on his brother's advice and went back to Mogilev. When he left, Alexandra, whom he told everything, was well aware of the anxiety that was beginning to take hold of him:

My precious! With anguish and anxiety I parted from you ... With all my heart I feel and suffer with you ... Our dear Friend prays for you also, in the other world ... It looks as though the situation is improving. But, dear one, you must be firm, show a strong hand, that's what Russians need. You *never* missed an occasion to show your love and kindness; let them now feel, for once, your fist ... I bless you, embrace you hard and press your tired head to my breast ... Feel my arms around you, my lips tenderly pressed to yours – together for ever, always inseparable.

Nicholas had returned to the front at Alexeyev's request. The Tsar was not far from Pskov when the 'February Days' began in Petrograd. The revolution had moved faster than the conspiracy, but at their crossing-point the same men were present – Rodzianko, who warned the Tsar of the dangers threatening the government; Guchkov and Milyukov, who were soon to see Michael and offer him the crown; and the generals, who, thinking that the conspiracy had outpaced the popular revolt, got the Tsar to abdicate. Two days later, they learned that Michael, too, had abdicated.

Thus, when the February Revolution took place, the Tsar had just arrived at Mogilev. The government in place was made up of men chosen by Alexandra. The chairman of the Council of Ministers, who had not long been in office, was Prince N. Golitsyn, a man without any political experience, whom Alexandra had met and admired as one of the organizers of the army's hospital services. He had declined the appointment, arguing that he was not up to it. 'If someone else had used the language I used to describe myself, I should have been obliged to challenge him to a duel.' But he had been forced to comply.

The 'strong man' of the government was Protopopov – well born, a protégé of Rasputin and a frequenter of seances. The *starets* had cured him of a form of syphilis that this sprightly deputy had contracted somewhere in Petersburg 'society'. The man who had been accused, to the Tsar's great astonishment, of having gone mad once he had become a minister was now really mad. He had discarded his deputy's jacket and put on the uniform of a general of the gendarmerie, on the pretext that he was now Minister of the Interior. In the Duma, where there were a dozen deputies who had been political exiles, this action was thought to be not in the best of taste. Rodzianko had mentioned Protopopov's name to the Tsar when there was question of forming a government 'of confidence', and Rasputin, too, had whispered his name to the Tsaritsa. With these recommendations, Protopopov had become a minister – but under Stürmer, which Rodzianko saw as a betrayal, since he had intended that Protopopov should serve under *him*.

The new minister replied: 'How can you ask me to resign? All my life it was my dream to be a governor, and here I am, a minister.'

Hated henceforth by the Octobrists and the Cadets, not to mention the socialists, and despised by the Right, Protopopov arrived at the

Duma with an icon that he carried ostentatiously and that, at his desk, he loudly consulted before addressing the deputies. He was heard to say: 'I feel that I am going to save Russia. There is nobody but me who can do it.'

The third representative of the Tsar's authority in Petrograd was the commander of the military forces in the capital. In theory this officer was under the orders of General Ruzsky, commanding the northern army group; but, knowing how she and Vyrubova were hated at Pskov, Alexandra had caused the independent military district of Petrograd, abolished in 1914, to be restored and placed directly under the Tsar. On Rasputin's advice she had caused this command to be given to General Khabalov. But when the Tsar ordered General Gurko to send Khabalov a reinforcement of two reserve divisions, Gurko, rather curiously, sent only two battalions from Pskov.

'The house is breaking up,' observed Zinaida Hippius. It was breaking from above and also from below, where discontent was acute. According to a police report dated at the beginning of 1917,

the proletariat here are on the verge of despair. It is thought that the slightest explosion, due to the least of pretexts, will result in uncontrollable riots ... The cost of living, which has trebled, the impossibility of finding food, the loss of time through queuing for hours outside shops, the rising death rate, the restrictions imposed on the workers, have become unbearable. The ban on moving from one factory to another or changing one's job has reduced the workers to the level of cattle fit only to serve as cannon-fodder. The ban on all meetings, even meetings held to form co-operatives or to organize canteens, and the ban on trade unions are causing the workers to adopt an openly hostile attitude towards the government.'

By contagion, this popular discontent affected the troops – the rear garrisons first, then the soldiers at the front, who were already angered by the heavy losses suffered in 1915 and 1916, for which they blamed their officers. In soldiers' letters intercepted by the censorship there was talk of a settlement of accounts when the war was over – 'or even sooner'.

Towards the middle of February stocks of flour in Petrograd fell to the lowest level. The district commander, General Khabalov, decided to introduce rationing. The people learned of this and, the very next day, queues formed when the bakeries opened their doors, a phenomenon soon to be repeated at all food shops. Sold out in a few hours, some of these shops then lowered their metal blinds. Crowds gathered, shops were broken into, and such incidents recurred in the days that followed. They burst out when, after hours of waiting in a temperature of less than 20 degrees, the crowd heard the fateful word 'netu' ('no more left').

The Duma had just begun to meet on 14 February and several deputies denounced the ministers as 'incompetent'. They called on them to resign, adding that in France the people had once 'swept away the throne'. But the new chairman of the council, Prince Golitsyn, and his ministers left the deputies to talk to themselves: to show their disdain they let the government benches remain unoccupied.

As they sensed the coming storm, the left-wing deputies sought contact with the illegal organizations. At Maxim Gorky's home deputies such as Kerensky and the Bolshevik representative Shlyapnikov met to discuss the situation. However, they could not agree. Some believed in taking action, others did not. There was only an exchange of sharp words between 'defencists' and 'internationalists'.[2]

At the same time, the socialist parties and the trade unions tried to prepare for a demonstration on 23 February (8 March, NS), Working Women's Day. Since, however, they did not manage to agree, the women decided to act on their own.

When, on the morning of the 23rd, these women workers formed a procession, followed by a certain number of men workers, the organizations called on everyone to join in. On that first day the

2. When war broke out the old political categories (populists, Marxists, and so on) became blurred, as from then on the conflict was between those who advocated defence of the fatherland, first and foremost (the 'defencists'), and those who wanted to carry out an 'international' action against the war.

women's demonstration was swollen by an influx of men who had been laid off by the Putilov works and then, later, by thousands of others. Fearing disturbances in the centre of the city, the authorities had ordered that offices and shops be shut. The office and shop employees, having thus been told not to come to work, watched the demonstration and then, along with many persons moved by curiosity, joined it. 'The strikers were grave and dignified,' an observer wrote. The petty bourgeois of St Petersburg were joining the workers in a protest against the Tsarist regime. For the first time in Russian history, the working class had emerged from its ghetto, and other social groups were showing their sympathy with it.

The mood was quite cheerful. 'It was like a festival day.' The trams stopped. The people made gestures of friendship to the Cossacks who were patrolling the streets. The passivity of the police amazed everyone.

On the second day, 24 February, the women workers were still playing the chief role. They meant to march along the Nevsky Prospekt, drawing as many people as possible into their ranks. By 8 a.m. the men workers had joined the women and they all began moving from the proletarian suburbs towards the centre of the city. This time, though, the police had taken up positions to prevent the demonstrators from crossing the bridges over the Neva. Ignoring the bridges, the demonstrators crossed over the ice and re-formed their ranks on the other bank. With red flags at their head they sang the 'Marseillaise'. A huge crowd was now gathered at the end of the Nevsky Prospekt, on Znamenskaya Square (now October Square). There were shouts of 'Up the Republic!' The Cossacks cantered by and were greeted. A demonstrator noticed that one of them winked at the crowd. But then the mounted police came on the scene, shouted 'Disperse!', drew their weapons, charged and wounded some of the demonstrators. But, lacking orders, they failed to pursue the fleeing people. Once more, the attitude of the Cossacks had been surprising.

On 25 February, the third day, the Bolsheviks were the chief organizers of the strikes and processions. Strikes had broken out again on an extraordinary scale.

The War Minister, Belyaev, had ordered that an attempt was

again to be made to stop the demonstrators crossing the Neva, but there must be no shooting, 'on account of the impression that would make on our Allies; but break the ice before they get there'. Nevertheless, General Khabalov issued no special orders and, as before, the people from the poor parts of the city were able to reach the centre. It was there, in Znamenskaya Square, that an incident occurred. A speaker was addressing the demonstrators when mounted policemen arrived and called on the crowd to disperse, but nobody moved. One of the policemen aimed his weapon at the speaker, the crowd began to shout, and amid a cloud of snow and dust a Cossack rode forward and cut down the 'pharaoh' (the people's name for one of these enforcers of law and order). The crowd stood flabbergasted, not knowing what to think.

That evening the discussion in the Council of Ministers was stormy. The Minister of the Interior was furious because the chairman had met Rodzianko without him, and shouted: 'I'll arrest your Rodzianko and dissolve the Duma.' But the most important event at this meeting was the arrival of General Khabalov with a telegram he had just received from the Tsar: 'I order that the disorders in the capital, intolerable during these difficult times of war with Germany and Austria, be ended tomorrow. Nicholas.'

Subsequently, testifying before the commission set up by the Provisional Government to examine these events, General Khabalov explained:

This telegram – how shall I put it? – well, to be frank, it was like a blow on the head for me. How could I end the disorders 'tomorrow'? What was I supposed to do? When the people asked for bread they were given bread, and that was the end of it. But when the banners were inscribed 'Down with the autocracy!' the people were not going to be pacified with bread. But what was to be done, then? The Tsar had given his orders. It was necessary to use force.

The fourth day was a Sunday. The people of St Petersburg got up later than usual. Once up, though, they found soldiers posted at key points. General Khabalov had already sent a telegram to the Tsar:

'Today, 26 February, calm has prevailed in the city since dawn.'

By midday the working-class parts of the city were in motion, and, in the centre, the streets were thronged. Soldiers blocked the avenues and surveyed the sidewalks. They were given their orders by bugle-calls sounded from the rear. But the crowd went up to the soldiers and spoke to them in a friendly way, and they replied in the same spirit. The officers kept issuing orders, so as to interrupt this dialogue. Those in command were irritated and tense as they felt their authority failing.

In the Duma, deputy V. Maklakov proposed a 'plan': simultaneous resignation of the government, suspension of the Duma for three days and formation of a new government 'of confidence', to be headed by some popular general like Alexeyev. By a government 'of confidence' was meant one that would be responsible to the Duma. However, the government rejected this proposal and decreed a state of siege. Sure that it had the situation in hand, it told the Tsar that there would be no 'fifth day'.

In fact, by the evening of the 26th the demonstrators were tired and discouraged. The political organizations were pessimistic and everybody thought that the revolution had failed again. 'In the early hours of the 27th,' wrote Trotsky,

the workers thought the solution of the problem of the insurrection infinitely more distant than it really was. It would be truer to say that they saw the problem as almost entirely ahead of them, when it was really nine-tenths behind. The revolutionary pressure of the workers on the barracks fell in with the existing revolutionary movement of the soldiers to the streets.

Angered by the order to fire that their officers had given them the previous day, the soldiers took over from the workers, locked up their officers and killed some of them, and joined the demonstrators. The soldiers' column merged with that of the workers and they all advanced on the Taurida Palace, where the Duma was sitting. The revolution had won.

Within a few hours, on 27 February, the Tsarist regime collapsed.

Symbolic of this defeat was the advance of another column of soldiers who marched, colours flying, to the Winter Palace. A French witness, the Comte de Chambrun, tells what happened:

While the Palace of Justice was burning, the Pavlovsky Regiment marched from its quarters, with its band playing. I watched these battalions pass in close order, led by non-commissioned officers. Instinctively, I followed them. To my surprise, they marched towards the Winter Palace, went in, saluted by the sentries, and invaded and occupied it. I waited a few moments and saw the imperial flag come down slowly, drawn by invisible hands. Soon after, alone on this snow-clad square, my heart heavy, I saw a red rag floating over the palace.

When 20,000 demonstrators entered the gardens of the Taurida Palace the deputies lost their heads. Some, fearful of being massacred, went out into the street to lose themselves in the crowd. Others, like Milyukov, thought it more dignified to stay put and resist any attack. The monarchist deputy Shulgin describes very well the anxiety of deputies who were uncertain whether the demonstrators had come to protect them or to attack them: 'Alarmed, disturbed, they huddled together in spirit, so to speak. Even persons who had been hostile for many years to the autocracy now suddenly felt that there was something that was a threat to them all alike, something dreadful and horrible. That something was the street, the mob.'

On the one hand, these Duma deputies, sensing the rise in the tide of popular anger, had continually urged the monarch to accept their help in order to resist the common foe – both the external, German foe and the internal one, the revolutionaries. On the other hand, so as to exert pressure on the regime, they had also continually attacked the government and the Tsar, and thus kept alive the discontent of the masses. The man of the moment was a socialist deputy, Alexander Kerensky, who rushed to meet the demonstrators and welcomed them in the name of the Duma. In his presence the Duma performed its first revolutionary act, by setting up, without permission, a 'Committee for restoring order and establishing relations with institutions and personalities', the very name of which constituted a programme.

At the same time, in a wing of the palace, some other socialist deputies were forming, together with leaders of the illegal organizations, a 'pre-Soviet' made up of representatives of the trade unions, the co-operative movement, the SR, Menshevik and Bolshevik parties and the anarchist groups. The Menshevik Chkheidze was elected chairman, and Kerensky, as deputy chairman, shuttled between the two committees. Next day a Provisional Government was formed, its legitimacy ensured by the Soviet. This government contained Prince Lvov, from the Union of Zemstvos, as chairman, together with Guchkov, Milyukov and Kerensky. Rodzianko was not a member; as Speaker of the Duma he set out to discover what was going on at GHQ and, above all, to discover what the Tsar might do.

The capital had certainly been won, 100 per cent, for the revolution, and nobody knew what had become of the old government. One witness relates that when, during the night of the 27th, amid the fusillade of shots, light came on again in the Marynsky Palace, the seat of the government, War Minister Belyaev was found hiding under a table. Everyone shared this fear, and it was especially acute in the case of those who had formed the Soviet or the Provisional Government. They all had in mind the precedent of 1848 in Vienna, when the Austrian Emperor had allowed the revolutionaries to become masters of the capital so as the more easily to surround it and crush them within its limits. In Petrograd it was fear that united all these 'class enemies' – fear of the crowd that was still there, finger on the trigger and trembling as much as they did, and fear, above all, of the pitiless repression that might follow. All sorts of rumours circulated. General Alexeyev had been made chairman of the Council of Ministers. The Tsar was going to come and put down the revolt on the spot. Grand Duke Nicholas was in supreme command again, and held the Peter-and-Paul Fortress. Everyone wondered how the soldiers at the front would react and what the Tsar would try to do.

'We are done for, done for!' cried Grinevich, a trade unionist member of the Soviet, in the hearing of Sukhanov, a fellow-member.

'It's the scaffold for us now,' thought Peshekhonov, another member of the Soviet, as he mounted the steps of the Taurida Palace.

The Tsar, having arrived at Mogilev on Thursday 23 February,

Working Women's Day, knew nothing about what had happened on that day. He answered the letter that Alexandra had left for him at the time of his departure:

You write that I should be firm and authoritative, and that is absolutely right. You can be sure that I don't forget it; but it is not at all necessary to be snapping at people, left and right, every minute of the day. A calm but sharp observation or reply is very often quite enough to make someone realize what his place is ... Here at Mogilev my mind is at rest. *No ministers, no problems.* But I suffer from our separation, I hate it. I shall not stay here long.

On the 24th, the second day of the revolution, he was still ignorant of the demonstrations taking place, but he learned of them on the 25th, from his wife's letter sent on the 24th, in which she told him about Kerensky's speech in the Duma: 'I hope that Kedrinsky [*sic*] ... will be hanged for his frightful speech. That's what's needed [martial law in wartime].'

Thus, GHQ and the Tsar had been officially informed of the troubles in Petrograd. General Khabalov had sent Alexeyev details of the first disturbances, and Protopopov, the Minister of the Interior, had wired to Voyeikov, the commander of the palace guard, who travelled with the Tsar and lived in the same carriage, that the movement was of a chaotic nature, 'vigorous measures having been taken to suppress the disorders'. The General Staff and the Tsar reacted in the same way. The commander of the northern army group was ordered to do everything possible to hasten the dispatch of reliable troops. 'Our future depends on it.' And Nicholas sent Khabalov his order that the 'intolerable disorders' must be 'ended tomorrow'.

This message, which had had such an effect on General Khabalov, did not trouble the Tsar much. 'The sovereign seems worried, but today he was cheerful,' noted General Lukomsky. In his diary for 25 and 26 February the Tsar made no mention of the events, and on 26 February he wrote to his wife: 'I hope Khabalov will be able to stop these street disorders. Protopopov must give him clear and definite instructions. If only old Golitsyn does not lose his head!' The illness of his children, who were suffering from influenza, concerned him much more, and he told Alexandra that he would be back at Tsarskoe Selo in two days' time.

In Petrograd, Sunday had passed in unexpected fashion. Until midday everything was still calm. The General Staff and the Tsar received reassuring reports. At about eleven, however, the Tsaritsa told Nicholas that she was worried. That afternoon there was shooting, and Rodzianko, alarmed and unaware of the Tsar's orders to Khabalov, sent Nicholas his first telegram, which provoked the recipient to remark: 'Here's fat Rodzianko writing some nonsense to me. I shan't even reply.'

That Sunday, the 26th, passed 'peacefully', according to General Dubensky. The Tsar continued to read Julius Caesar's *Gallic War* and took afternoon tea as usual. Next morning, the 27th, he received further reassurance from Belyaev and Khabalov. While reporting the mutinies that had occurred during the night, they both asked for urgent reinforcements, but the ruthless measures mentioned by the War Minister were not such as to worry Nicholas II.

When, consequently, on the 27th, the fifth day, the Duma Committee and the Petrograd Soviet came into existence together, the Tsar's position was clear. Two days earlier he had not wished to listen to Generals Ruzsky and Brusilov when they urged him to form a government in co-operation with Rodzianko. On the 27th he appointed General Ivanov 'dictator' and announced his departure for Tsarskoe Selo. He had, therefore, decided in favour of repressive measures. At the same time he told his brother Michael that he was not going to delay his return to Tsarskoe Selo, that he would not change his ministers unless he saw fit to do so after he had been to Petrograd, that General Ivanov would see to the suppression of the disorders and that reliable troops were on their way.

But none of those things happened. General Ivanov did not manage to approach the capital; the General Staff ceased to send him supplies, in the belief that the Duma had the situation in Petrograd in hand. Nor did the Tsar manage to reach Tsarskoe Selo, the railway workers having taken up the rails on the line (as also on Ivanov's route). He had to go back to Pskov, where he learned that the army and the government were now asking him to abdicate.

At the moment when the capital was wholly in the hands of the rebels, the Tsar was still having talks with General Ivanov, who was telling him how he put down a mutiny at Harbin in 1906. On another occasion, at Kronstadt, some sailors had been fighting among themselves: Ivanov had taken a couple of them by the scruff of the neck and ordered them to get down on their knees. The sailors obeyed him and the brawl ceased, to the amazed admiration of everyone present.

General Ivanov was confident he could enter Petrograd without bloodshed. 'But of course,' the Tsar replied, as he went off to bed at three in the morning. During that night of 27–28 February he had shown himself, according to witnesses, 'kindly, mild, quiet'.

Throughout 28 February, Generalissimo Alexeyev, unaware of the turn taken by events in the capital – the formation of a provisional government by agreement with the Petrograd Soviet – was informing all the army commanders of the preceding developments, describing the rebels in the capital as mutineers, and reminding each one of his duty to the sovereign. The result was that, at the moment when the revolution had triumphed completely, the imperial train was speeding towards Tsarskoe Selo in complete ignorance of what had occurred. There was at this time no difference of view between the Tsar and his General Staff. Though not really uninformed, Nicholas was not thinking of a political solution to the troubles. Only military measures had been taken.

The 'dictator' left Mogilev on the morning of the 28th. He intended to proceed by train to Tsarskoe Selo, with an élite battalion made up of soldiers who had been awarded the Cross of St George. On his arrival he would study the situation along with the Tsaritsa, while awaiting the reinforcements that were due to join him from Pskov or Reval (Tallinn). The first 'incident' happened at Dno station, where General Ivanov's train met one that had arrived from Petrograd. Civilians and soldiers told the troops of the 'dictator' about the extraordinary events that had just been happening in the capital. In a flash, the General's train was abandoned, and his men gathered on the opposite platform. From his carriage General Ivanov heard such words as 'all equal' and 'freedom'. He tried to regain control over his men, but they laughed at him. The 'dictator' rushed up to the nearest group and ordered them to kneel, putting them under arrest. But a whistle blew and the train moved off, taking Ivanov's soldiers to freedom.

Further along the line it was the railway workers who held up the General's train. 'The line's damaged,' they said. Ivanov made them repair it, but that took several hours. When he got to Tsarskoe Selo a disappointment awaited him. The St George regiment, thought to be the most loyal, told him that, in the event of conflict with the inhabitants of Petrograd who had gone over to the revolution, they would remain neutral, since their oath obliged them to defend only 'the Tsar's person' – and the Tsar was not there. The regiment then hoisted a white flag. Ivanov, in agreement with the Tsaritsa, decided that it would be wisest to pull back and wait for the fresh troops he needed.

Instead of those troops, Ivanov received two telegrams, one from Alexeyev at Mogilev, the other from Nicholas at Pskov. They both ordered him to suspend operations and await the arrival of the Tsar. He learned also that Guchkov, representing the Provisional Government, would come to meet him somewhere along his route. Lacking both troops and instructions, the 'dictator' did not know what to do next. The railwaymen had taken up the sleepers behind his train so that no extra forces could reach him. When he got to Semrino the 'dictator' lost patience and called for a fresh locomotive. Did it take so long to fill up with water? Never mind. Since he wanted a locomotive without delay, the railwaymen gave him one, chosen at random. An hour later, a little beyond Semrino, Ivanov's train came to a halt, owing to lack of water. It was not to resume its journey, and so Ivanov's adventure ended.

Faced with the worsening of the situation, on that same day, 27 February, War Minister Belyaev had, before disappearing, sent a final telegram to General Alexeyev, with a copy to General Ruzsky, commanding the northern army group. 'The situation in Petrograd has become very grave. The few loyal units are no longer capable of putting down the revolt, and more and more of them are going over to the rebels ... It is essential that reliable units be sent here in sufficient numbers.' At the same time Speaker Rodzianko sent another alarming telegram to General Ruzsky. He drew to his attention the food shortage, the strikes and the general paralysis of normal life. 'If order be not restored, shame and humiliation threaten Russia, as we shall be unable to continue the war under these conditions. The only solution is to entrust the charge of forming a government to a man who enjoys the country's confidence ... We cannot wait – waiting means death.'

In this document there was no more mention of 'responsibility to the Duma', and that for good reason: in Petrograd the Duma could no longer act without the approval of the Soviet. As soon as he received these telegrams, General Ruzsky telegraphed to the Tsar to tell him of his anxiety regarding the disorders. Pointing out that the army represented all classes of society and, on that ground, venturing to express a view, he concluded: 'Repressive measures would merely aggravate the situation.'

At the same moment the Tsar received a telegram signed by twenty-two members of the State Council, including Count Tolstoy and Prince Trubetskoy, who 'respectfully' urged him 'to alter fundamentally his domestic policy and choose a Prime Minister who enjoyed the country's confidence'.

Faced with these multiple and convergent pressures, Nicholas now sent off a telegram ordering General Ivanov 'to do nothing until he arrived at Tsarskoe Selo'. But he did not know that GHQ had done the same, halting Ivanov's advance because they thought the Duma had the situation in Petrograd under control. Stopping Ivanov meant, implicitly, that they recognized the Duma as the new ruling power.

While this was going on, Nicholas was on his way to Tsarskoe Selo. On 28 February, at about 3 p.m., he had telegraphed to his wife: 'Wonderful weather. I hope that you are feeling well and are calm. Many troops have been sent from the front. Heartiest greetings. Nicky.' He learned of the formation of the Duma Committee presided over by Rodzianko – which had taken place two days earlier –

but nevertheless confirmed to the Tsaritsa that he would be home 'tomorrow morning'.

During 1 March 'there was no talk of the events' in the imperial train. 'How shameful,' Nicholas noted merely in his diary; 'it is impossible to reach my palace at Tsarskoe Selo, but my heart and thoughts are there all the time. How painful it must be for my poor Alix to live through these events all alone.'

As for the Tsaritsa, she was amazed at the failure of the Tsar's uncle Paul to prevent disloyal behaviour in the Guards under his command. 'I am astonished. What is Paul doing? He ought to have kept them in hand.' But Paul had preferred, in concert with the other Grand Dukes, to compose a manifesto to be presented to the Tsar.

We propose that a new regime be established as soon as the war is over ... It is a matter of very great sadness to us to perceive that while Russia's fate is being decided on the battlefield, an agitation is diverting the people from their work, which is so badly needed if the war is to be carried through to victory ... Confident in the aid of Providence, we are sure that, for the sake of our country's well-being, the Russian people will pull themselves together and will not allow the enemy's intrigues to succeed.

The Grand Dukes wanted the Tsar to issue a proclamation on these lines:

Under the sign of the Cross we grant the Russian state a constitutional regime and order that the work of the State Council and of the Duma, which was interrupted by our decree, be resumed. We instruct the Speaker of the Duma to form immediately a provisional government and, in agreement with us, to convoke a legislative assembly which will without delay examine a draft constitution to be put before it by the government.

This was signed by Grand Dukes Paul, Michael and Cyril, and dated 1 March 1917. In his memoirs Paul claims that the Tsaritsa approved the Grand Dukes' move and the document they had composed. This is not probable. She sent to Nicholas this note, which he may have received: 'Paul wants to save us by a method which at once is noble and senseless. He has drawn up a stupid manifesto about a constitution after the war, and so forth.' She also wrote him another letter, her last as Tsaritsa, but which did not reach him: 'Everything is against us, and events are unfolding at lightning speed. It is plain, however, that they don't want to let me see you, so that they can get you to sign some paper, a constitution or something. And you are alone, without your army behind you, caught like a mouse in a trap: what can you do?'

At 8 p.m. on 1 March the Tsar at last arrived at Pskov. General Ruzsky, commanding the northern army group, was waiting impatiently on the station platform. All day long telegrams from Rodzianko, Alexeyev and Ivanov had been piling up, each requiring an urgent reply.

'Let us have dinner first,' said the Tsar.

General Ruzsky had been struck by the fact that persons like Grand Duke Sergei Mikhailovich, General Brusilov and Major-General Hanbury-Williams, head of the British military mission, were all advising the Tsar to grant Russia a responsible government and a constitution. He had been struck even more forcibly by the revolt in Moscow and the rallying of Admiral Nepenin to the revolution. Convinced that Nicholas had already lost too much time, he became more pressing. He realized that the Tsar and his circle failed to appreciate the scale of the disaster and that they mainly blamed General Khabalov for not managing to put down the disorders. Fortunately there was still Ivanov, to whom instructions would be sent.

On reading the telegrams reporting revolution in Moscow, in the Baltic Fleet and at Kronstadt, the Tsar received a shock. Nevertheless, when Ruzsky spoke to him about a responsible government, the Tsar rejected the suggestion – 'calmly, coolly, but with profound conviction'. 'Those ministers, as soon as they have failed, will walk away and wash their hands ... I will not go against my conscience,' he said.

At eleven a telegram arrived from General Alexeyev, also calling on the Tsar to form a responsible government. Now, at last, Nicholas gave in – although, in the manifesto he issued, he reserved for himself the right to appoint the Ministers of War and Foreign Affairs.

General Ruzsky returned to the attack when he received a call from Rodzianko. The General at once told the Speaker that the Tsar was going to appoint him Prime Minister, but Rodzianko replied: 'His Majesty and yourself are apparently unable to realize what is happening in the capital ... The question of the throne will have to be faced.' Rodzianko had been present at the meeting between the Duma Committee and the Soviet at which the Provisional Government had been set up with the Soviet's backing. He explained to Ruzsky that, even with amendments, the manifesto no longer met the requirements of the situation. Abdication was needed if the country was to be saved from anarchy. In the light of all this, Ruzsky decided that it would be best not to issue the manifesto.

At his end, Alexeyev had learned of the mutiny of the battalions at Luga, when they refused to submit to the orders of 'dictator' Ivanov. At 10.15 the chief of staff contacted his senior commanders to tell them of the conversations he had had, and invited them, accordingly, to urge the Tsar to abdicate, 'so as to preserve the country's independence and save the dynasty'. All the Generals replied by the deadline given them, except Evert. In anxious mood, Alexeyev asked to see the Tsar again. Nicholas received him in company with Count Freedericks, General Danilov and General Savich. He had with him the telegrams from his Generals advocating abdication, and also another telegram, from Grand Duke Nicholas Nikolaevich.

The Tsar smoked cigarette after cigarette. The atmosphere was tense. Then, suddenly, he said: 'I have taken my decision. I shall abdicate in favour of Alexei.' So saying, he crossed himself. He turned to General Ruzsky, embraced him and thanked him for his services. In the document to be sent to Rodzianko he asked to be allowed to remain with his son until Alexei should come to the throne.

General Voyeikov, Admiral Nilov and Count Freedericks rushed in at this moment, furious and intending to arrest Ruzsky. But Nicholas showed them the telegrams. 'What else could I do? They have all betrayed me, even Nikolasha.'

He consulted his doctor concerning his son's actual state of health, and received confirmation that the boy had little chance of living much longer. Then he altered the terms of his abdication, naming his brother Michael as his successor.

Once more, however, an unexpected event hindered publication of the announcement of abdication. Guchkov and Shulgin arrived at Pskov as delegates from the Duma. The men round the Tsar, especially General Voyeikov, recovered hope for a moment, but were disappointed. The visitors had come on behalf of the Duma to receive the Tsar's abdication. Like the military, they considered that swift action was needed, 'in the interest of the dynasty'.

'We bowed,' writes Shulgin:

The Emperor greeted us, shaking hands. This gesture was rather friendly ... The Emperor took a place on one side of the small square table that was placed along the green silk wall. On the other side of the table sat Guchkov. I sat near Guchkov, obliquely across from the Emperor. In front of the Emperor was Count Freedericks.

Guchkov was speaking. He was greatly moved. He evidently spoke well-thought-out words, but it was difficult for him to master his emotion. He did not speak fluently and his voice was dull.

The Emperor ... looked straight ahead. His face was absolutely calm and impenetrable. My eyes did not leave his face ...

Guchkov spoke of what was happening in Petrograd ... He spoke (as was his habit) covering his forehead slightly with his hand, as if to concentrate better. He did not look at the Emperor and spoke as if he were addressing some inner person ...

He told the truth. He did not exaggerate, nor did he conceal anything. He spoke of what we had seen in Petrograd. He could not say anything else. We did not know what was happening elsewhere. We were crushed by Petrograd, not by Russia.

The Emperor looked straight in front of him, calmly; he was completely impenetrable. The only thing that it seemed to me could be guessed from his face was: 'This long speech is superfluous.'

At that moment General Ruzsky came in. He bowed to the Emperor and, without interrupting Guchkov, sat down between Count Freedericks and myself ...

Guchkov showed his feelings again, as he broached the subject that perhaps the only way out of this situation was for the Tsar to abdicate ...

... The Emperor replied. After Guchkov's emotional words, his were calm, simple and precise. Only his accent was slightly strange – the accent of a Guards officer. 'I have decided to abdicate. Until three o'clock yesterday I thought that I could abdicate in favour of my son Alexei. But I have now changed my decision in favour of my brother Michael. I hope you will understand the feelings of a father.' The last sentence he spoke more softly.

When the text of the abdication document had been checked, the Tsar left for Mogilev. On the station platform the officers present tried to hold back their tears. Nicholas saluted them and briskly boarded the train. In his diary he merely noted: 'Left Pskov at one in the morning with a heavy heart: all around me I see treason, cowardice and deceit.'

A few days later, Anna Vyrubova told her mother, in a letter, all that the Tsar ('Papa') had told her about his abdication:[3]

'Weary, tormented by disturbing rumours and trembling for my family and especially for Mama and the little one,' he told me, 'I feared, and I still fear, more than anything else, that Mama will fall victim to hatred. And can the little one be saved? This thought affected me to such a point

3. When it first appeared, in 1928, and then again at the beginning of the 1950s, Anna Vyrubova's *Freylina eya Velichestva, Intimny dnevnik i vospominaniya* (published in Paris by Payot in 1928 in a translation entitled *Journal Secret 1909–1917*) was subjected to doubts as to its authenticity. The same suspicion had not applied to her *Souvenirs de ma vie* (Payot, 1927).

Actually, the revelations of the *Journal Secret* were bound to upset monarchist circles, for they give a far from flattering picture of the coterie around the imperial couple, and, above all, their tone is sharp, caustic, bitter and crudely different from that of the earlier book, which was tearful and full of kind sentiments, so that one may wonder if it was indeed the same person who wrote such contrasting works. It has to be remembered, however, that the *Journal Secret* was not meant for publication, and we do not know what part was played by the editor, S. Karachevtsev. All the same, the substance of the information given here corroborates evidence from other sources. Did Vyrubova, like Tartuffe, have two faces, perhaps?

that I could find rest only in prayer. At that moment Shulgin and Guchkov arrived. At the mere sight of Guchkov I knew that this would be a terrible blow. Shulgin was the one to speak. His voice quivered and there were tears in his eyes. He said that I should abdicate in favour of the little one, who would be helped by Michael. I did not get the point at first. Then I understood the main thing – that the little one was alive. With an effort I asked: "Is my family safe and sound?" Guchkov answered, fear in his eyes: "Yes, Sire, and, so long as we can, we shall do everything to protect your family."' Papa left the room, saying that he wanted a few minutes to reflect. The salutary idea struck him that Baby ought to be sheltered from harm. And then, the most terrible thought, that they would separate Baby from Papa and Mama. That might be the death of him. Better, then, to remove him at once from all danger. This was why he decided to abdicate for Baby as well as for himself. He said, too: 'This was a decision forced on me, a knife was raised to strike me, and I could always prove that, if asked. And, anyway, it would mean a rest for us. All together. Away from everything. No more suffering, no more fear.'

Kissing the cross given him by the *starets*, he remembered his words: 'Always do what comes into your mind after you have received a blow. The first thought is for your salvation, it comes from God. What comes later is the Devil's work.'

'Remembering that, I crossed myself, for Baby and for myself. They were very abashed. Guchkov said to me: "Sire, you are more a father than a Tsar."''

These words admirably describe the character of a ruler who was, in a sense, two persons – the autocrat and the human being – and who, by abdicating, made plain the basic contradiction that had led to the revolution.

Embodying a principle, autocracy, that he thought it his duty to perpetuate, Nicholas II had no alternative but to resign when he was forced to grant something that his 'conscience' forbade him to grant. Establishing a government in accordance with popular representation would be contrary to the very principle of autocracy. On that point he had been uncompromising ever since his accession, and the concessions he had had to make had always been wrested from him, so that he tried to retract them from time to time. There was no exception to the continuity of his conduct; in 1906, in 1911 and again in 1915 he had no intention of sharing his authority with anyone, being ready, at most, to listen to opinions. He would have found it intolerable, given his coronation oath, if even that much had been imposed on him as an obligation. He did not want a Duma, and all this supervision, all this dialoguing with the government, were for Nicholas an enfringement of the prerogatives that God had conferred upon him.

Consequently, he suffered when men who had been associated with him in that destiny abandoned him in his determination. He had shown towards Alexei the concern that he would have wished his own father had shown towards him, by teaching him his duties as Tsar, instead of leaving him to know only the outward forms of Tsardom while remaining incompetent for his task. And now the only thing that he thought he knew how to do, to be a good father, had been made futile because his son was doomed. The tragedy of this man was that he had regarded it as his duty to transmit a power 'intact' to his son, and that he now realized that no such transmission would take place. So he abandoned this power that had overwhelmed, burdened and vexed him, without having changed his view of it, and completely unconscious of the crimes he had committed in the name of the autocracy.

Films have preserved the memory of the extraordinary joy that came over Russia at the news that the Romanovs had abdicated: soldiers marching along, shooting in the air, self-appointed tribunes chanting slogans, masses of people happy and eager to have their say. Every Russian has in his pocket a plan all ready for putting Russia back on her feet. Rich and poor, coachmen and officers, men and women follow each other on to the platforms from which to harangue the crowd. The faces are lit up with excitement and radiate happiness. Images of joy and renewal. On 23 March 1917 the funeral of those who fell in the cause of the revolution took place. The funeral march resounded at this first secular interment in Orthodox Russia: 'Farewell, brothers, you have come to the end of your path of glory ... '

A civil funeral, a meeting, free speech – these were the first signs of a world turned upside-down. Russia had become the freest of all the belligerent countries, Lenin was soon to say, and, indeed, in town and country alike, in church as in university, all traditional authorities were being flouted – where they had survived, that is.

Kerensky wrote about this period that there were 'three days without any government whatsoever'. Within a few days, however, there was no town or other community without some revolutionary organ, soviet or committee, which took the place of the traditional authorities. The old government had disappeared during the February days, along with the former provincial governors and the much-hated Okhrana. Apart from the acts of violence committed in the course of the revolutionary days, particularly against officers, only the policemen of Elizavetgrad, who had been responsible for a number of pogroms, provided the expiatory sacrifices of the February Revolution. The message they left before going out to face a firing-squad expresses well the atmosphere of the February days of 1917:

We, the policemen of Elizavetgrad, salute the soviet and the government and congratulate them on having caused the triumph of liberty, so long in coming. About to die, we prostrate ourselves before the Russian people and beg them to forgive all the evil that we unwittingly committed in the performance of our duties . . . It would have been sweet to die from enemy bullets, our souls at peace, our children proud of us.

'We fear neither God nor Devil now.' This statement made to their priest by some parishioners on the morrow of the abdication gives us a notion of how the muzhiks saw the nature of Tsardom. For them it was the expression of all their woes: Tsarism, but not the Tsar in person, for very few peasants, workers or soldiers demanded, in their petitions to the Petrograd Soviet, that measures be taken against Nicholas. What was challenged was the regime and, more than the Tsar himself, his henchmen. An army chaplain reported that his flock, some front-line soldiers, said to him: 'We don't trust you any more; you're a tool of the old order. Go away, that's the best thing for you to do if you don't want us to give you a beating.'

Another testimony of this period: 'We were covered with insults by Tsarism, and bound hand and foot,' said a peasant to his village priest. 'And you, who went along with it, you dazzled us and blinded us. All you knew was to sing "God Save the Tsar".' The Orthodox clergy were thus seen as agents of tyranny and did not enjoy that halo of beneficence that served the Catholic clergy well in France, for instance, during the revolutions of 1789 and 1848.

The officer corps did not realize, either, that the soldiers attributed the rules of discipline to the autocracy, so that it seemed to them self-evident that, once Tsarism had been overthrown, they would be changed. The soldiers did not contest military discipline as such, but the way it was used and the excessive forms it assumed. They saw it as a guarantee of the social order and believed that the officers, by refusing to change the rules, were trying, in their own way, to perpetuate Tsarism without the Tsar. This was one of the reasons for the mutinies of spring 1917. They were not due to unwillingness to fight, except in so far as the men thought their commanders were conducting offensives as a means of getting back control of the troops – thanks, precisely, to that discipline which became legitimate on the battlefield because it ensured proper conduct by the soldiers.

In the countryside the heads of households assembled at once when the news of the abdication reached them. The village commune revived, so to speak, spontaneously. After dictating to the local lawyer or teacher the wishes they formulated for themselves and for the well-being of the country, they waited for Petrograd to reply. Since no reply came, they occupied uncultivated land, in particular land belonging to the Tsar and the apanages. Why should they wait for a decision by the Constituent Assembly? It was right that these lands should be returned to them, and the Tsar's abdication meant that justice would at last be done.

Hatred and resentment against officer, priest and landlord, rather than against the person of Nicholas II, were manifested when the Tsar was moved from place to place, and particularly at Tobolsk. True, in that relatively prosperous region there had never been large-scale strikes or repressions. It was in the poor quarters of the cities, mainly, that he was called Nicholas 'the Bloody.' Even in those working-class districts, though, nobody demanded 'vengeance' as the Tsar passed through during the February Revolution, at any rate. It was necessary for the ruling classes to resist reforms or to sabotage the economy, and then for the forces of counter-revolution to lift their heads, before talk about the Tsar's guilt could become transformed into hostile actions and the imperial family made victims of violence.

As for the symbolism of the Tsar's person, that vanished, and the new regime replaced the ceremonies of the old one with new rituals: solemn funerals, demonstrations, mass processions.

The symbolic power of Tsarism was such that, when Nicholas II fell, everything that had been associated with him was automatically discredited. Exemplary from this angle was the fate suffered by the Duma. It had never ceased demanding, and Nicholas had never ceased ignoring it. But that was already too much. After February, the Duma faded away. Its deputies no longer dared to appear as such in public, or to meet together, except for one or two occasions, in the course of several months. On the day of its seeming triumph, when it formed itself into a committee and then into a Provisional Government, the Duma proclaimed its end. In a sense, it committed suicide, owing to the fact that it constituted a link between past and present.

Never had Lenin, never had any of the revolutionaries from Kerensky to Trotsky or Kropotkin, appealed to the Tsar. Never had they negotiated with Tsarism; never had they recognized it. They alone could take over now.

Some, like Milyukov, himself an historian, felt very quickly that this was so. By sacrificing Nicholas's person they sought to save what it had incarnated.

When, during the night of 1 March, negotiations ended in the Taurida Palace between the delegates of the Duma Committee and those of the Petrograd Soviet, one of the new ministers, Milyukov, named his colleagues: Kerensky for Minister of Justice, Prince Lvov for chairman, and so on. When he was asked for the Cabinet's programme, Milyukov began to expound this, mentioning that it was the product of an agreement with the Soviet, but he was interrupted by shouts of 'And the dynasty? What about the Romanovs?'

The Cadet leader replied: 'I know that my answer will not satisfy all of you here, but I shall give it. The old despot who has brought the country to utter ruin will abdicate the throne or be deposed. The power will pass to a regent, the Grand Duke Michael Alexandrovich. Alexei will succeed him.'

Shouts of 'Down with the dynasty!' interrupted Milyukov at this point. When he was able to resume, he said:

Gentlemen, you don't like the old dynasty. I may not like it either. But now the question is not who likes what. We cannot leave the question of the constitution of the state without a decision ... As soon as the danger is past and order is restored, we shall proceed to the convocation of a Constituent Assembly, elected on the basis of a universal, equal and secret ballot. A freely elected popular representative body will decide who more faithfully expresses Russian public opinion – we, or our opponents.

Actually, Milyukov was lying. Abolition of the monarchy, demanded by some deputies in the Soviet, had been rejected at his *own* request and that of other deputies. Milyukov had pointed out that, until universal suffrage spoke, a decision like this would lack legitimacy. Provisionally, at least, he had won on this point.

But Milyukov, Rodzianko and the rest would soon have to retreat. The government made it known, in effect, that Milyukov had expressed only his personal views on the fate of the dynasty. The new ministers and members of the Duma felt once more the breath of civil war. The deputies knew very well, and were frightened by the knowledge, that it was vitally important for the population not to learn that Michael was to succeed Nicholas. But the news reached Moscow, and at once hundreds of telegrams of protest were sent to the Soviet.

Two telegrams were given to Guchkov and Shulgin as they got out of their train; it was feared that, unaware of the state of opinion in the capital, they would make public the information they had brought back from Pskov. 'It would be best to say nothing more about the Romanovs until the Constituent Assembly meets,' Rodzianko explained. 'If people in the capital were to learn that the

Duma and GHQ were agreed on recognizing Michael II, there would be no answering for the consequences.' Alexeyev and Ruzsky let themselves be convinced and tried to recall their telegram. Paris and London knew of it, all the same – but not Petrograd.

Meanwhile, Guchkov and Shulgin, when they arrived at Petrograd station, thought they were bringing good news when they announced that Nicholas II had abdicated and Michael II would be his successor; they narrowly escaped lynching. 'Down with the Romanovs – Nicholas, Michael, they're all the same! Down with autocracy!' The government's agents came along in time, reassured the crowd, relieved Guchkov of his abdication document and saved the situation.

Warned by the turn events were taking, Michael at once realized that he was not cast in the heroic mould. To the delegates of the Duma who called on him he put one question only: 'Can you guarantee my life if I should take the crown?'

Thereafter the outcome of the conversation between Rodzianko, Kerensky and Michael was a foregone conclusion, and Milyukov's chidings were useless. A lawyer, Nabokov, drafted the act of abdication in such a way as to leave open the possibility of a restoration of the monarchy (a concession to Milyukov and Guchkov), and the government decided to publish simultaneously the two renunciations, that of Nicholas and that of Michael. Greatly moved, the last Romanov agreed to this procedure and signed without hesitation.

When the news was made public, there was rejoicing all over the capital. There were gatherings again, at which people shouted and sang. 'Now it's all over,' said a member of the Soviet to a friend who was with him in the midst of the delirious crowd. He heard a woman standing nearby say softly: 'You're wrong there, *batushka*; not enough blood has been shed yet.'

Actually, no one expected a revolution so rapid as this, a change so violent, and the General Staff least of all. When he announced the formation of a new government, Grand Duke Nicholas concealed from his soldiers the nature of the upheaval and ended his address by calling on the men to sing 'God Save the Tsar'. He also warned them, resorting to the expressions that were most familiar to him, 'that any attempt to disobey the government's orders would be dealt with with the full force of the law'. The Romanov family counted on Nicholas Nikolaevich to 'take the situation in hand', as we know from a letter from Maria Pavlovna, Nicholas II's aunt, to her son Boris; and the victors of February might well fear the use he could make of the forces under his command in Caucasia. The Soviet and Kerensky succeeded in getting the Grand Duke dismissed, together with General Evert, who had refused to come to terms with the new regime.

General Alexeyev, whom the court suspected of wanting to profit by the circumstances to make himself dictator, behaved equivocally. He announced that Grand Duke Nicholas had been appointed commander-in-chief at the wish of the Tsar, and that the new government had been formed as a result of an agreement between the Duma and the Senate, which he knew to be untrue. As for General Brusilov, he said that these events had been accomplished by the will of God. Tomorrow just as today, the soldiers must carry out their sacred duty towards him. 'May the Lord help us to save Holy Russia,' he concluded.

Thus, the High Command of the army behaved as though a palace revolution had succeeded, thanks to intervention by 'the street'.

Disorders continued, and this annoyed Alexeyev. He wished to receive no more alarmist messages. For his part, General Kornilov, who was reputed to be a republican, announced his appointment as commander of the Petrograd district using the formula favoured by the Duma: 'Since the old regime had shown itself unable to cope, a new government took power.' Both generals strove to ensure that the soldiers should believe that this change, initiated from above, would in no way affect the established order.

There had been murders in the capital, when some officers were shot. It was in the navy, however, that most horrors were committed. Some mutineers of 1905 were still in prison in the fortresses of Reval and Helsingfors, and the sailors' hatred of their officers exceeded anything to be observed elsewhere. Also, the commanders of the navy remained loyal to Nicholas II to a greater degree than their counterparts in the army. They preferred death to abjuration.

No dialogue was possible between the rebellious sailors and their officers. Blood flowed and forty officers lost their lives, among them Admiral Nepenin, even though he had declared his support for the new regime. At Kronstadt, Admiral Wiren died bravely: 'I have spent my life in loyal service to my Tsar and my country. I am ready. It is your turn now. Try to give a meaning to your lives.' When he was shot he wanted to fall forward.

The more that men had enjoyed the Tsar's favour, the more quickly did they go over to the new regime. Grand Duke Cyril set the example, followed by the Cossacks of the Guards, the palace police and His Majesty's own regiment. Few stayed loyal: Grand Duke Paul, the only one of the family to remain with Alexandra, the Novgorod Horse Guards, Count Keller, the Benckendorffs and Count Zamoyski, who went on foot to Tsarskoe Selo to offer his services to the Tsaritsa. Bunting, the governor of Tver, committed suicide. Although the former Minister of Supply, Bark, had many friends in the new government, he declined to stay in office, being offered the portfolio of Finance. 'It's a question of principle,' he said.

After prolonged slumber, Grand Dukes and Generals who owed everything to their Tsar had abandoned him with light hearts. When, a few months later, they wanted to go back on their decision, it was too late.

The journey of Bublikov, who was entrusted by the Petrograd Soviet with the task of conducting the Tsar from Mogilev to Tsarskoe Selo, was a triumphal progress.

At each station the local people came in crowds to greet us, making speeches and shouting 'Hurrah!' I replied likewise. At Mogilev too we were received with a general 'Hurrah!' The delegate of the Soviet asked General Alexeyev to inform the ex-Tsar that he was now a prisoner. Not daring to put it so directly, he used a periphrasis that Nicholas was to repeat to everyone he met: 'Be it known to you that, henceforth, the Tsar is deprived of liberty.'

Alexandra knew nothing of all this. Not being able to communicate with the Tsar from Tsarskoe Selo, she had gathered round her such members of her entourage as seemed to her the most loyal, namely, Protopopov, who wanted to appeal to the shade of Rasputin, and Paul, 'who was convinced that it was now pointless to appeal to his soldiers, since they would support the Provisional Government'. She had written to Nicholas, when still unaware that he had abdicated: 'And you are alone, without your army behind you, caught like a mouse in a trap: what can you do? This is the greatest baseness and meanness, without precedent in history ... If they force you to make concessions you are in no case obliged to fulfil them, since they were obtained from you by unworthy methods.'

On the following day (3 March) she wrote: 'Paul has just been here and has told me everything ... It could make us go mad, but we won't ... You will be crowned by God himself on this earth, in your own country.' In this letter she referred to 'Judas Ruzsky'.

The Tsar had asked General Alexeyev not to put any obstacle in the way of his return to Tsarskoe Selo, and had also asked to be allowed to go into exile until the end of the war and then to come back and settle permanently in the Crimea. The Provisional Government agreed to these requests.

Anna Vyrubova has left an account of the Tsar's return to the bosom of his family:

At present he is thinking of nothing beyond the protection of his family. His wife is always with him, caring for him as though he were ill. Not a word of reproach. He does not complain of anything, but I learn from Fyodorov [the doctor at GHQ] that he has had a bad heart attack. He is now trying to recover. The people around him are in a terrible state, as though stupefied. They are suffering to the point of desolation, for him above all. But they do not complain. Once he said: 'If I were not a Christian, the simplest thing would be to put an end to things straight away ... Perhaps that would be the best solution for the family ... ' I tried to persuade him that if he did that, its effect, besides the impairment of his dignity, would be to kill his wife. 'That is what I am afraid of.' And then, with a sad smile, he gave his word that he would think no more of that.

Kerensky, in his role as the Minister of Justice, was unwilling to be 'the Marat of the Russian Revolution'. If he had opened the prisons, it was not so as to put a fresh lot of 'delinquents' into them. True, there were policemen and members of the Okhrana who had plenty of crimes on their consciences. But, in a burst of generosity, the revolutionary of February 1917 decided to treat these guilty men as innocent. 'What have you done, why are you here?' a warder demanded of a prisoner on the eve of the revolution. 'I am here so that *you* may no longer be struck or humiliated,' was the prisoner's reply. And in February 1917 the prisoner freed the warder from his infamous task.

Kerensky arranged with Milyukov, the Minister of Foreign Affairs, that, 'at the request of the Provisional Government', the British government would agree to the departure of the imperial family for England, where they would be welcomed 'for humanitarian reasons'. Actually, this deed of benevolence would also provide Britain with a guarantee that reactionary circles would not attempt a restoration with one or other of Nicholas's heirs who might fix up a separate peace with Germany, something that, in this spring of 1917, was a danger much on Lloyd George's mind.

The British Prime Minister acted, therefore, with determination; and as soon as his agreement had been received, through Ambassador Buchanan, on 2 March 1917, the Provisional Government offered Nicholas the choice – either to go to England or to remain in Russia.

It has been said, though there is no satisfactory evidence for this, that the ex-Tsar had already agreed to accept an invitation secretly sent by his cousin Willy. In order to save his wife and children he would have been ready, if the story is true, to put himself in the hands of the enemy. However, this operation is supposed to have failed – once again through action by railway workers on strike, who, without knowing it, prevented the Tsar from reaching the Gulf of Finland. However that may be, Nicholas did, we know, accept Kerensky's proposal that he go to England, and Major-General Hanbury-Williams was informed accordingly. On 6 March the route to be taken, via Murmansk, was decided upon, and Nicholas recorded in his diary that he was packing his bags.

On 9 March, however, having learned that the Provisional Government had offered Nicholas Romanov the possibility of leaving for England, and knowing that he was on his way to Petrograd, the Executive Committee of the Soviet decided 'to take immediate measures to keep him under arrest'. This was the resolution adopted by the committee on that day:

An order has been sent out warning all railway stations, so that our armed detachments may check on the passage of trains. A special commissar has been appointed with full powers over the railway stations of Tsarskoe Selo, Tosno and Zvanka [this was Mstislavsky]. Telegrams have been sent to all stations, for Nicholas's arrest. The Provisional Government has been warned of the Soviet's opposition to Nicholas Romanov's departure for England, and it has been decided to transfer him to the Peter-and-Paul Fortress, the garrison of which has accordingly been changed.

Further on in the minutes of this session, the following information is given regarding the situation at Tsarskoe Selo:

The palace guard is well controlled by revolutionary detachments. An order has provided that nobody is to enter or leave. The telephone lines have been cut. Nicholas is under close surveillance. There are about 300 soldiers there to guard him, from the Third Rifle Regiment. Permission to remain in the palace has been given to Dolgoruki and the Benckendorffs. Letters and telegrams are being intercepted ... It is reported that there is some agitation against the Soviet in the First Reserve Regiment, stationed at Tsarskoe Selo, with the troops calling for a constitutional monarchy. The officers are responsible for this agitation.

In fact, after the Petrograd Soviet's emissary Mstislavsky had been able to see the ex-Tsar, in his palace of Tsarskoe Selo, with his wife and children, the Soviet fell in with Kerensky's arguments against transferring the family to a fortress. Like a romantic hero, the spokesman of the February Revolution was unwilling to let blood be shed. He had already saved the lives of the ex-ministers by protecting them with his own body from infuriated rioters. Surreptitiously, they were given passports so that they could escape abroad. He also wished to save the lives of the ex-Tsar and his wife, which could be managed more easily from Tsarskoe Selo than from a fortress standing in the midst of the working-class districts of the capital.

A few weeks later, however, the situation was completely altered, and it was now not the Provisional Government who had problems but the ex-Tsar himself.

In England a press campaign prepared for the coming of the ex-Tsar and his family, recalling 'his loyalty to his Allies'. When the Petrograd Soviet learned of this, they demanded explanations. Wasn't Nicholas going to be put on trial, as Louis XVI had been, and as the ex-Tsar's ministers were to be? Milyukov at once told the British that transferring the family would cause trouble, and Kerensky advised them that they should not press him too obviously if they wanted the ex-Tsar to be able to leave in good time. It is worth recalling that Milyukov called himself a monarchist and Kerensky a republican. Lloyd George had to deal with a powerful movement of opposition to his project. The Labour Party and the trade unions protested against Nicholas 'the Bloody' and his wife, the German Empress, being allowed to come to Britain. There was the prospect of a general strike, in connection with the extraordinary news that was coming from Russia, which aroused amazing enthusiasm in trade union and pacifist circles. The Tsar's arrival in Britain would be seen as a provocation. The government recoiled at once, and when asked in Parliament if they had made any proposal regarding the Tsar's future domicile Lord Robert Cecil replied that 'His Majesty's Government have taken no initiative in this matter' – which, taken literally, was not untrue. Faced with the scale of the protest and the rising tide of strikes, King George V intervened in person to request that the government withdraw its proposal to grant asylum to the Romanovs.

In Russia it was Ambassador Buchanan who had the delicate task of informing the Provisional Government that Britain had changed its mind. Later, Meriel Buchanan, the ambassador's daughter, was to reveal that her father was threatened with loss of his pension if he made public the role played by the monarch in this turn-about.

In the meantime the strikes had ceased, the Tsar had been murdered and, at home, Buchanan was being blamed for what had happened.

But Nicholas now knew nothing of governments and their policies, or wished to know nothing. It is said that when he was told of his cousin's change of attitude he remarked: 'I feel shame for Georgie and for his country.'

One of the most painful experiences for the Tsar was when, in the presence of his son Alexei, he received an order from Kerensky and had to obey it, like a subordinate. On another occasion, when he was walking in the grounds of the palace in mid-April, the sentry on duty blocked his path and said: 'You can't pass, Colonel.' The officer who was with the Tsar and his son intervened, but Alexei went very red when he saw the soldier compelling his father to halt.

Nicholas's diary is the same in the post-abdication period as when he was still on the throne. On 9 June:

It is just three months since I returned from Mogilev and we have been held prisoner here. How painful it is to be without news of my dear mother. Nothing else matters. It is even warmer today than yesterday (20 degrees in the shade, and 36 in the sun). Again a smell of burning. After my walk I gave Alexei a history lesson. We worked well, in the same place as yesterday. Alexandra did not come out. Before dinner the five of us went for a walk.

The ex-Tsar also notes down the books he has been reading. In April, Uspensky's *History of the Byzantine Empire*, 'begun 4 April, ended 25 April, Tuesday, 870 large pages'. Then *The Valley of Fear* by Conan Doyle, and *A Millionaire Girl*, finished on the 28th. In the meantime he had begun Kasso's *Russia on the Danube* and also General Kuropatkin's *Problems of the Russian Army*. In the evening he read aloud to his family *The Mystery of the Yellow Room*, *The Perfume of the Lady in Black*, *The Haunted Armchair* and *The Luck of the Vails*, followed by Merezhkovsky's *Alexander I*.

There is only one allusion to the outside world, on 22 May: 'Today is the anniversary of the offensive of our forces on the south-western front ... What our mood was then, and what a difference today.'

On 19 July (OS), the anniversary of Germany's declaration of war on Russia: 'O Lord, come to our aid and save Russia!' After which Nicholas drew a cross.

Alexei's tutor, Pierre Gilliard, who stayed to the end with the imperial family, tells us that the ex-Tsar's sole consolation during his forced residence at Tsarskoe Selo was reading the French newspapers. The republican newspapers, of course. He expected no less, one may suppose, from the monarchist press which, with *L'Action Française*, wrote that the Tsar had abdicated 'of his own free will, in order to save the people from a revolution'. Alexandra repeated this argument several times during their stay in the 'gilded cage' of Tsarskoe Selo. Whenever she thought she had a grievance against the regime of surveillance, which became more and more rigorous, she would say that 'the Tsar abdicated in order to spare the people's blood'. Some British newspapers claimed that 'this revolution was neither anti-dynastic nor anti-aristocratic but merely anti-German'. However, the British were not as warm as the French towards the Tsar, 'the most loyal of the Allies'. 'We must begin by saluting the noble sovereign whose name remains associated with the alliance between our two countries,' wrote Alfred Capus in *Le Figaro*, and later in *Le Gaulois*. These extracts from the press appeared in the *Journal des débats*, which Nicholas read and which also quoted articles in which it was stated that 'no measure has been taken against the Romanov dynasty'. It was indeed the case that, until the Constituent Assembly met, abolition of the monarchy was, perhaps, only something wished for, but not yet a reality.

That was, of course, going to change, and soon *Le Canard enchaîné* did not fail to mock at all who were turning their coats. 'Yesterday, the Tsar was not to be touched. Today ... ' And a caricature showed Nicholas, barefoot in prison, with the caption: 'Ah, if only Gustave [Hervé] could come and set me free!' Hervé, a 'patriotic' socialist, had written at the time of the abdication a lyrical and enthusiastic article that the entire press had reprinted. Not a word said against the ex-Tsar. True, 'the autocracy has been overthrown but Slavism stands ... The cause of the Allies has won here its finest victory. The army will at last have behind it an administration that is modern, honest and patriotic ... What a body-blow for the Kaiser ... What an example for the German people.'

Everything that had happened was due, according to Hervé, to the hatred felt by the Russians for the Germanophil bureaucracy of Protopopov and company, but *his* reign was now ended. Nicholas II emerged untouched from these exaggerations. Proof of this is to be seen in the fact that, when it became known that he was under arrest, Hervé and all the enthusiasts for the complete overturn began to get worried. 'May our Russian friends permit us to say to them, in all frankness, that the arrest of the Tsar has made us "wince" a little. When that road is taken, one knows where one begins but not where one will end up.'

He had begun to say what, without admitting it publicly, most of the leaders of France – Ribot, Painlevé, Pétain – were afraid of: that disorder in Russia would hinder the country's war effort; that the revolution would compromise the planned offensive by the Russian Army; and – who could say? – one day, perhaps, the Russians might make a separate peace.

The tone of the French press deceived its readers to the extent that it offered, even to the ex-Tsar, a very serious reassurance.

When Kerensky endeavoured to save the life of Nicholas II he was moved by humane considerations, certainly, but also by a certain idea he had of the revolution. It must, above all, avoid bloodshed, put an end to humiliations and refrain from turning executioners into victims and victims into torturers. He was not the only one to cherish that ideal, but he embodied it more strongly and with more conviction than the other revolutionaries. And this was the reason for his popularity. Like all the Russian socialists, he was a republican, and, on this question, the fate of the Tsar, no one has ever suspected him of having manoeuvred for political ends.

That was not the case with all of the principal actors in the February Revolution. In the Provisional Government, Milyukov, Minister of Foreign Affairs and leader of the Cadets, did indeed have ulterior motives. Like Prince Lvov and some others he was a monarchist; and although, like all his party, he was against autocracy and wanted a constitution to serve as a check on absolutism, he certainly did not want the democracy of the soviets, those councils of workers and soldiers, nor even a republic. This had been clearly seen on the night of 1–2 March, when the pact between the Duma Committee and the Executive Committee of the Petrograd Soviet was negotiated, that episode of sharp bargaining at the end of a revolutionary week that left soldiers and workers encamped in the Taurida Palace. For the Soviet to agree to the formation of a Provisional Government it was necessary to yield on two of its demands: for an amnesty and for political freedom. On the third point – not predetermining the nature of the future regime, as between republic and monarchy – Milyukov was immovable. Sukhanov, a member of the Executive Committee of the Soviet, tells us:

What I personally found rather unexpected was not that Milyukov should fight for the Romanov monarchy but that he should make this the most contentious point of all our terms. *Now* [in 1920] I understand him very well and think that from his point of view he was absolutely right and extremely perspicacious. He was calculating that with a Romanov Tsar, and perhaps *only* with him, he would win the forthcoming battle ... He thought that, given a Romanov Tsar, all the rest would follow, and was not afraid – not *so* afraid – either of the freedom of the army or of 'some' Constituent Assembly.

Milyukov and Guchkov were furious at the abdication of Michael II. It was not the person of Nicholas II or that of Michael II that mattered to them so much as what they personified. However, nothing irreversible had been done, since a republic had not been proclaimed and, thanks to Nabokov, who had drafted Michael's act of abdication, the possibility of a restoration of the monarchy had been safeguarded. But who would be the heir to the throne? Olga, sister of Nicholas and Michael? Olga, the ex-Tsar's eldest daughter? Or Cyril, in the most direct line, the son of Vladimir, Alexander III's brother and the cousin of Nicholas and Michael?

The leaders of the army considered that they had been duped. Would they have obliged the Tsar to abdicate if they had realized the extent of the revolution that had taken place in Petrograd? The messages that passed from Rodzianko to Ruzsky, to the effect that the Duma Committee had the situation well in hand, had convinced the Generals, who had long been prepared for this transfer of power and who now misunderstood the course being taken by events. The proclamation of 'Order No. 1', on the rights of soldiers, suddenly revealed to them the magnitude of the disaster. War Minister Guchkov and Prince Lvov wielded power only in appearance. They governed *to the extent that* the Petrograd Soviet approved of what they were doing. The Tsar had only been gone a fortnight when already the Soviet was talking of 'peace without annexations or indemnities'.

The new government was, to be sure, resisting the notion of a 'peace without annexations or indemnities' put forward by the Petrograd Soviet; but, from the standpoint of the General Staff, what a road had been travelled! They had joined in a plot to overthrow the Tsar so as to prevent negotiations with Germany and wage the war more effectively – and now, hardly had Nicholas fallen than people were talking about peace. The April crisis, with the fall of Guchkov and Milyukov, showed the degree to which the revolutionary movement had become radicalized. Above all, with the Bolshevik propaganda that exposed the annexationist ambitions of the Provisional Government (in the name of respect for treaties signed), with the fraternization at the front, with the questioning of the planned offensive, not to mention the growth of workers' demands and actions by peasant committees, the new head of the government, Kerensky, was showing himself incapable of stabilizing the situation.

For the free citizens of spring 1917 in Russia, everything representing the state was being rejected as a survival from Tsardom. In the first place, the Provisional Government itself was suspect as an institution that had issued, more or less, from the Duma. To it they opposed the Soviet, even though the leaders of the Soviet, who were socialists, became ministers in April. In its turn, every town opposed its own soviet to the Soviet in the capital. Legally, they had the same rights, so why should not the town's soviet run the town's own affairs? But then, in its turn, every collectivity – factory committee, district committee, and so on – opposed itself, the 'base', to the soviet of deputies. Instances were actually set up in opposition to the policy of the Executive Committees, and, in a way, it was the representative system that was being challenged; the revolution meant direct government by the governed themselves. Any delegation of power was disapproved of, any 'authority' was felt to be unbearable. The framework of the state burst asunder and the propaganda of both Bolsheviks and anarchists encouraged this tendency.

For the big bourgeoisie and the army leaders, Russia was hastening towards catastrophe, Russia was in her death-throes. The dual power must be ended, the soviets got rid of, the Bolsheviks arrested, their leaders shot. The idea of appealing to a Cavaignac came naturally to them, and so it was around the officers' leagues created in May that this idea made progress.

An 'anti-revolutionary model' took shape that had something in common with the Fascist model that later arose in Italy: defensive reaction against social revolution; action by the military and the church; challenge to the class struggle; appeal to the manly solidarity of the fighting men; recourse to special action groups like commandos; appearance of new men, often former revolutionaries who had rallied to the cause of national defence, like Boris Savinkov; anti-Semitism; establishment of 'cells' in the state apparatus; use of violence against democratic institutions; denunciation of the weakness of the country's leaders; sympathy and intervention on the part of the Allied governments. Not a word was spoken about the monarchy, and about the Romanovs even less. Minds had evolved so far in six months that the men who came together to restore, so to speak, Tsardom without a Tsar called their organization the Republican Centre.

Kerensky had dismissed General Alexeyev, who was too closely linked with the old regime, and then General Brusilov, who, ready to accept reforms, had lost the confidence of his colleagues. His place was taken by General Kornilov, a 'republican' general. He was a republican who dreamt only of a showdown with the soviets. He was the candidate of the Republican Centre, a candidate for the role of dictator.

'I have good hope that he will succeed,' said Alexandra to her husband when, through Anna Vyrubova and monarchist circles, she learned about what was going on. 'I, too,' said Nicholas.

At the beginning of summer 1917 Kerensky tried to take the army in hand and restored the death penalty at the front with a view to re-establishing discipline – though this penalty was never applied. 'Let us hope that this measure has not been taken too late,' was the ex-Tsar's comment.

For some weeks the monarchists, having recovered their confidence, had been trying to mount an operation that would rescue the ex-Tsar and his family. 'May the Holy Icon be blest if he agrees to a revolt by the empire,' said N. Markov, who was close to Anna Vyrubova. In reply, Alexandra and Nicholas sent him an icon of St Nicholas the Wonder-Worker. But that Markov was unable to accomplish anything, any more than S. Markov, connected with the Republican Centre, who seems to have plotted a rescue attempt. This group sought support abroad: in London, in Madrid, in Nice and in Lausanne, where the first leaflets calling for Nicholas's return were printed. In the Crimea the Dowager Empress, Grand Duke Nicholas and Grand Duke Michael also organized an action group comprising Countess Keller, Baron Korff and Senator Taneyev, Anna Vyrubova's father. Tracts headed *Vpered za Tsarya i Svyatuyu Rus* ('Forward, for Tsar and Holy Russia') were distributed by them in Yalta.

With the July insurrection, which went beyond control by the Bolsheviks themselves and aimed to overthrow the Lvov–Kerensky government, the threat that hung over the imperial family grew greater. The accusations that Lenin and the Bolsheviks were paid agents of Germany won over some of the soldiers and caused the insurrection to fail. Kerensky, now alone in power, decided that the imperial family should be sent somewhere out of reach of another popular flare-up, as unpredictable as the 'July days' had been. A monarchist coup was also, in his view, not out of the question. These reasons argued for exile in some more distant place. Not in the Crimea, where public opinion would not easily agree to the ex-Tsar resuming residence in a palace, and where the sailors, far from reliable, might as probably execute him as liberate him, but in western Siberia, at Tobolsk, for example, remote from the dangers of revolution and reaction alike. The Soviet historian G. Z. Ioffe has found the telegram that determined the departure of the family for their far-off destination: 'Special train, destination known to you, will leave 31 July, arriving 3 August Tyumen. Prepare for that day boat to go to Tobolsk, and arrange accommodation accordingly'.

An escort of 330 soldiers was to ensure the security of the move, which was to take place under the flag of the Japanese Red Cross. The government allowed Michael to embrace Nicholas before he set off for his exile. Kerensky was present. In order to let them converse freely, he said: 'I shall shut my ears and not listen. Go ahead.' And he turned towards the window-recess, leaving them to themselves.

Tobolsk had not experienced the October insurrection. Or, more precisely, Bolshevik power was established there only much later. If a date must be given, 15 April 1918 would doubtless be the most exact. But the reality was more complex. The different parts of the empire did not evolve in the same way. As regards radicalization, it was Petrograd that began the advance, and it ended in Tobolsk.

Hardly had Kerensky dispatched Nicholas and his family far from the storms he sensed were approaching than there occurred the *putsch* by General Kornilov, which he had not wanted to foresee, having himself placed Kornilov at the head of the army.

The counter-stroke to Kornilov came at once. It came from Kerensky himself, who appealed to the population, but was based essentially on the network of soviets and revolutionary committees of all kinds that had been formed in the capital since the February Revolution. After the 'July days' and the repression that followed, the Bolsheviks had sunk back underground. Lenin had gone into hiding, but the Bolshevik workers and soldiers were both surprised and indignant that their party should have been charged with being an agent of Germany. With the Kornilov *putsch* and the appeal 'to all' by the Menshevik–SR government, the Bolsheviks emerged from clandestinity. They had foreseen this military reaction, and as the analyses and forecasts of their leaders were shown to be correct, their strength in the soviets grew tenfold. In Petrograd one of them, the former Menshevik Trotsky, was elected chairman of the Soviet.

Kornilov's defeat was the work of all the socialist parties, but it nevertheless weakened Kerensky because, thereafter, he could no longer rely on the moderate elements and the Cadets to resist the Bolshevik upsurge. Moreover, the network of soviets, as they became more radical, was starting to constitute a sort of parallel state that as yet had no head (in October the Bolshevik Party would step into that role), while the state itself, contrariwise, certainly had a head, the Provisional Government, but its organs no longer carried out its orders, because the soviets controlled the country's life. In Petrograd, in some other towns as well and in the army, the Bolsheviks now held the majority in soviets of all kinds. They were thus able to take power and consolidate their position through an armed insurrection that set the seal on their mastery of the country.

But there were a number of regions, Tobolsk for instance, where the leadership of the local soviet was not Bolshevik but Menshevik and SR. That was the situation when Nicholas was taken to Tobolsk with his family, and it continued so even after October. In Tobolsk power was actually wielded by a security committee in which the Bolsheviks were very much in the minority. V. Pankratov, appointed by Kerensky to guard the ex-Tsar, provided Nicholas and his family with a tolerant regime that allowed them to go for walks in the town and to be visited by priests – to such an extent that the soldiers guarding them felt isolated and as though prisoners themselves. These soldiers informed Petrograd accordingly, pointing out that such conditions would make escape possible. 'We shall die but they shan't get out of it alive,' they told the provisional revolutionary committee headed by Sverdlov. They secured the recall of V. Pankratov and his replacement by a commissar named Pignatti; but this did not make much difference, and the Ural Bolsheviks, who were stronger in towns other than Tobolsk, found it incomprehensible that so many links could be established, without any reaction by the authorities, between the ex-Tsar and visitors whose intentions seemed obvious.

The correspondence between the Tsaritsa and Anna Vyrubova testifies to the ex-Tsar's extreme passivity in face of all that was happening to him. We have a score of letters written by Alexandra at Tobolsk to her close friend, most of them in Russian, others in English and one in Old Slavonic. They are all filled with the same mysticism and express the same love.

'My dear martyr,' wrote the Tsaritsa to Anna, herself in prison:

I cannot write, my heart overflows, I love you, we love you, we thank you, bless you and bow to you. We kiss the wound on your forehead and your eyes, filled with suffering. I cannot find the right words, but you and I know everything, distance does not affect our love ... Our souls are united, and through suffering we understand each other still better. My family bless you and pray for you ceaselessly ... Love takes no account of distance, and in thoughts and prayers we are always together ...

We live here far from everybody and life is quiet, but we read of all the horrors that are going on. But I shall not speak of them. You live in their very centre and that is enough for you to bear ...

Nobody is now allowed to approach us ...

It is very cold – 24 degrees of frost. We shiver in the rooms, and there is always a strong draught from the windows ... We all have chilblains on our fingers ... Little Jimmy [a pet dog] lies near me while his mistress plays the piano. On the 6th Alexei, Marie and Gilik [M. Gilliard] acted a little play for us. The others are committing to memory scenes from French plays. Excellent distraction, and good for the memory. The evenings we spend together. He [Nicholas] reads aloud to us and I embroider ... I also have lessons with the children, as the priest is no longer permitted to come ... Do you read the Bible I gave you? Do you know that there is now a much more complete edition? I have given one to the children and I have managed to get a large one for myself. There are some beautiful passages in the Proverbs of Solomon. The Psalms also give me peace ...

He is amazing, with such strength of character, however much he is suffering. I am astonished to see him like this. All the other members of the family are brave and never complain.

In January 1918 the ex-Tsaritsa learned that Anna Vyrubova had met Gorky, and that he was proposing to intervene on her behalf. 'I was very surprised to learn that you had met Gorky. He used to be horrible, an abnormal creature, he wrote disgusting books and plays. Is it really the same man? He fought against Papa and Russia when he was in Italy. Be careful, my dear.' Next day: 'Be cautious with people who come to see you. I am very worried about Bitter [the English equivalent of *Gorky*]. He edits a frightful paper[4] ... Don't say anything serious to him ... Bitter is a real Bolshevik.' And she ends by mentioning that they have had a visit from a priest.

The coming of that priest and the links established with the Bishop of Tobolsk were to have consequences.

What was actually going on, and why did the Bolshevik authorities not react? In fact, they were not able to. For example, they could not transfer the Tsar to Kronstadt, as their most determined supporters demanded. They thought that nothing could happen until the ice on the rivers broke up, so that the Tsar was confined safely at least until the spring. On several occasions it was decided in Moscow that 'the problem of the ex-Tsar is not urgent'.

In reality, much activity was in progress among the monarchists, who proved able to collect large sums of money – over 200,000 roubles according to Benckendorff, and perhaps another 175,000 as well. The clergy of Tobolsk, headed by Bishop Hermogen, constituted a network of information on the spot through which contacts were made and the imperial family kept in the picture. In a letter to Vyrubova in January 1918 Alexandra expresses pleasure that 'Yaroshinsky has not forgotten us'. He had financed the hospitals of which Maria and Anastasia were patrons. In March the Tsaritsa voices fear concerning the fate of Boris Soloviev, who was Rasputin's son-in-law, and whom, of course, the family trusted; he was said to be preparing some scheme to rescue them. There was also a group of monarchists around ex-senator Tugan-Baranovsky; they had succeeded in renting a house opposite the governor's residence, which was where Nicholas and his family were lodged, and had begun digging a tunnel. However, their plan came to nothing because the Tsar was moved soon afterwards to Ekaterinburg.

---

4. The reference is to *Novaya Zhizn*, a left-wing Menshevik journal that was friendly to the Bolsheviks before October but hostile afterwards.

The principal escape operation, though, was the one being organized around a movement called the Right Centre (Pravy Tsentr), which was led by such people as Krivoshein, Gurko, A. Trepov, General Ivanov and those who, on 1 March 1917, had advocated the issuing of a manifesto for the establishment of a constitutional monarchy. Markov II was connected with them and made contact for them with Madame Vyrubova and the Tobolsk clergy. But a twofold split in their ranks prevented them from acting. Some of them regarded the abdication as null and void, whereas others accepted it, but were themselves divided on the issue of who – Michael, Paul or some other – should act as regent for young Alexei. Yet another split divided the supporters of the Entente among them from the pro-Germans.

When the peace treaty of Brest-Litovsk was signed the pro-Germans were in a better position to intrigue with the Kaiser's representatives who installed themselves in Moscow. Monarchists hastened to make contact with the Germans, proposing that the latter effect a restoration in return for economic collaboration, which they would ensure through their links, Yaroshinsky's especially, with the German banks. But Wilhelm II would not hear of this: what was of prime importance was his arrangement with the Bolsheviks whereby he could transfer troops from the Russian front to France. He told the King of Denmark that any help given to the Romanov family, even purely humanitarian in character, would look like support for a restoration and compromise the 'peace process'. In Moscow the Kaiser's envoy Count Mirbach was later to repeat this consideration to all those who, through the Right Centre, were trying to secure the liberation of the Romanovs. There could, for the time being, be no question of overthrowing the Bolsheviks. The best the Germans could do was to help the monarchists organize themselves in preparation for a coup to be effected when circumstances had changed.

When, despite opposition from the great majority of Russians, the Bolsheviks signed the treaty of Brest-Litovsk, this was followed, at the beginning of April 1918, by a shift in the balance of power in the executive committee of the Ural district soviet, with the

Bolsheviks now in the ascendant there. They dissolved the local Duma and zemstvos and made themselves the effective rulers of the region. Although there was no apparent connection between these events, they both affected the fate of the imperial family. At a time when the monarchists were hoping that the Germans would help them bring about a restoration, the Bolsheviks of the Ural district decided that they must put an end to the permanent threat of escape by the Tsar. Commissar Pignatti on the spot and Sverdlov in Moscow considered that the problem of what to do with the Romanovs was now urgent. On 1 April the presidium of the Central Executive Committee resolved to move the imperial family from Tobolsk to another place.

Now there occurred the first of the unexplained events in this story, the mission of V. Yakovlev, who had been associated with Mstislavsky in the confinement of the ex-Tsar at Tsarskoe Selo. In early April this representative of Sverdlov, accompanied by thirty soldiers with four machine-guns, passed through Ufa and Ekaterinburg on his way to Tobolsk, with the mission of transferring the Romanov family to some other place. He was not allowed to say where, and claimed not to know anyway; instructions were to be passed to him at each stage of his journey. However, various contretemps frustrated whatever plan there may have been. Young Alexei had a crisis in his haemophilia and could not be moved for the moment. Insecurity prevailed in the region, where the different cities were quarrelling over control of communications, stocks of goods, and so on, often resorting to arms, while bands of irregulars added to the disorder.

Departure from Tobolsk therefore took place in two groups, with the ex-Tsar leaving first and the Empress and the children following later. These movements aroused suspicion among the various authorities and other armed forces in the region, which resulted in the prisoners being taken this way and that, now towards Moscow and now towards Omsk, which at that time was half controlled by the SRs. In the end, and after five days of adventures, Yakovlev brought the Tsar and his family to Ekaterinburg and handed them over to Beloborodov, Goloshchekin and the other members of the executive committee of the Ural district soviet. The Tsar was to be imprisoned there in a closely guarded house, his final home.

Now, nothing is clear in this story, even though some features of it are well attested. First, the Tsar and his family were led to believe that they were going to Moscow, whence they would be taken to some port in Scandinavia or England. Yakovlev gave them to understand this, though without saying so explicitly. Second, when he arrived at Ekaterinburg, Yakovlev was able to show that he had been constantly in direct contact with Sverdlov (or with Lenin) throughout his journeyings, and had obeyed his instructions. Third, this mission could be seen as having failed owing to local conditions, which had prevented Yakovlev from getting to any place but Ekaterinburg, where, with Moscow's approval, he left the ex-imperial family in the hands of the soviet. Fourth, subsequently Yakovlev went into hiding, then tried to rejoin the camp of the SRs, his original political family.

The affair becomes obscure because the detour in the direction of Omsk, where anti-Bolsheviks were in control, remains partly unexplained. The reference made to Riga more than once in the documents suggests that the family's departure for Scandinavia would take place through a port controlled by the Germans. Furthermore, the Tsaritsa relates that, *en route*, she met Sedov, an emissary of N. Markov who was connected with B. Soloviev and other monarchists. For Soviet historians, Yakovlev was a traitor. But is the matter as simple as that?

Let us put forward a twofold hypothesis that may make a little more intelligible this badly explained and rather curious affair.

In April 1918 pressure on the Bolsheviks by the Germans was very severe, and just as Wilhelm II played a double game between the monarchists and the Bolsheviks, so Lenin, Trotsky and Sverdlov played a double game between the Germans and the Allies. The British landing at Murmansk was not originally just anti-Bolshevik in intention. Its purpose was also to prevent an extension of German power into Russia's far north and Finland. Trotsky realized this and arranged for these Allied troops, as a countervailing force to the Germans, to be welcomed by the local soviet. To let the Tsar get away through a northern port, via Moscow or otherwise, was not inconceivable, if he was indeed to be sent into exile abroad. At this stage nobody in Moscow had really raised the question of executing the ex-Tsar; at most, they had thought of putting him on trial. Was Yakovlev's mission connected with this project? And was he trying to save the Tsar's life by any means, even those that had not been thought of?

Or else, my second hypothesis, it was by agreement with the Germans that Sverdlov was going to send the Tsar to Riga, whence he would proceed to Scandinavia and from there to England – which would be less of a shock to public opinion (if the arrangement leaked out) and to the Tsar himself. Nicholas believed that the Soviet government wanted to make use of him in bargaining with the Germans, and repudiated that idea. He saw Yakovlev as an agent of the Bolsheviks and of the Germans. Perhaps he was not wrong; but he did not know about the links that existed, also, between the Germans and his own supporters!

Wherever the truth lies, we have no proof either way. And Yakovlev managed to disappear. No certainty and few clues exist. The most disquieting fact is that, from this point on, as investigation advances, the clues become more and more uncertain and unverifiable, and we find ourselves in a huge forest full of fog that prevents us from seeing clearly.

The ex-Tsar describes his arrival in Ekaterinburg:

17 April: Tuesday. Again a splendid, warm day. We arrived at Ekaterinburg at 8.40 and remained in a station for three hours. A lively discussion went on between the local commissars and those with us. Eventually the locals won the argument and our train was moved to another station, the one for goods trains ... We were led through deserted streets to the house that had been got ready for us, Ipatiev's house.

The house is good and clean ... Four large rooms have been set aside for us – a bedroom, in a corner of the house, a bathroom, a dining-room with windows overlooking a small garden and from which we can see the lower part of the town, and a very big sitting-room ... Our luggage was checked with extreme strictness, it was a real customs examination.

Everything was looked at, down to the last phial in Alix's travelling medicine-chest. This exasperated me and I told the commissar very curtly what I thought of it ...

18 April: We slept very well indeed ... We heard a procession with music go by, celebrating 1 May [Nicholas's diary is still dated according to the Old Style, thirteen days behind the Gregorian calendar]. We were not allowed to go out into the garden, but got some air through a ventilation pane in a window.

21 April: Holy Saturday ... We were given permission to receive a visit from a priest and a deacon. At eight they performed the Easter service, quickly and well. It was a great consolation to be able to pray, even under these conditions and to hear: 'Christ is risen!' ... The soldiers on guard attended the service.

1 May: ... At noon the guard was changed, the new one being a special group of front-line soldiers, Russian and Latvian ... We have been told that we are not allowed to walk outside for more than one hour each day 'so that the regime may be like in a prison'.

2 May: ... An old housepainter has whitewashed the windows of all our rooms ... Commissars watch us when we go out for exercise.

The diary becomes less and less dense. There is still, however, this entry for May:

28 May: A very warm day. In the shed they are opening our trunks, which have arrived from Tobolsk ... They are taking things out of them, which makes me think that they may very likely make off with whatever takes their fancy, and we shall never see it again. Disgusting. Their behaviour towards us has altered in the same way. Our gaolers try not to talk to us. I have a feeling that they are worried, afraid that something may happen ... I can't understand it.

21 June: We have a new commandant. Andreyev has been replaced by Yurovsky ... He has taken most of our jewels ...

23 June: Yurovsky brought our jewels back in a box that he sealed after asking us to check the contents, and then returned to us to look after ...

Yurovsky has begun to realize that the men guarding us keep for themselves most of the provisions that people send in for us ... I am on the seventh volume of the works of Saltykov-Shchedrin, which I like very much.

Finally, the last entry in the diary: '30 June: Alexei has had his first bath since Tobolsk. His knee is better, though he can't bend it completely yet. The weather is mild and pleasant. No news from outside.'

A few days later the ex-Tsar was executed.

4. An Event – or an Item for 'News in Brief': An Enigmatic Death

Nineteen eighty-nine. 'Moscow prays for the innocent Tsar.' 'The Russians must know the truth about the Tsar's death.' These headlines from Soviet newspapers, inconceivable only a few years back, tell us of the demand for truth in a society that now lives under the sign of *glasnost*! Any and every opportunity is taken for lifting part of the veil from a forbidden history, that of the beginnings of the Soviet regime. In April 1989 the writer Geli Ryabov announced in *Moscow News* that he had, ten years previously, found the remains of the Romanovs, or at least their skulls, showing that they had not been destroyed with acid but had been buried somewhere a long way from Ekaterinburg. The implication of this 'discovery' is obvious. If these really are the remains of the imperial family that have been found, it will be possible to transfer them, say a mass for them and honour their memory; this has already been stated by Father Vadim and by Vladimir Anichenko, one of the leaders of a committee formed to rehabilitate the former Tsar.[1]

This same Geli Ryabov published in May 1989, in numbers 4 and 5 of *Rodina*, a very long article quoting previously unpublished extracts from a confession by Yurovsky, the man in charge of the execution of the Romanovs and the commandant of the Ipatiev house, the 'house of special purpose' where the Romanovs were put to death. This document appears to have been written in July 1920, that is, two years after the killings. In the same issues of *Rodina* the historian G. Z. Ioffe, a specialist in the period, discusses at length what is known about the execution, and in particular the question of responsibility – local or central.

---

1. On 12 November 1989 an Orthodox Constitutional Monarchist Party was formed at a monastery near Moscow.

The need for truth, innocent or not, certainly accounts for the reappearance of testimonies and documents concerning those events; the Russians had been starved of them. But do these sensational articles say anything more than what we knew already, in the West at least? Can we be sure that the version of the execution of the Romanovs that we had in the West is the true one? The confessions of Yurovsky, for example, even if they are in the Soviets' public records – are they more reliable than other confessions such as history has known many of, in the USSR? In this enigma, testimonies and depositions are sometimes quite contradictory, though they are all in the records.

But what, actually, are the sources of information at our disposal? On the killings themselves the basic material was for a long time the report published by N. Sokolov, one of the judges entrusted in 1919 with the task of investigating what happened, after the Whites had discovered evidence of the crime. It was entitled *Judicial Inquiry into the Assassination of the Russian Imperial Family*.[2] Before this report appeared, Sokolov's superior, General Dieterichs, had published, in 1922, *Ubiystvo Tsarskoy Sem'i* ('The Murder of the Imperial Family'), the conclusions of which he had already given to the world, on the basis of Sokolov's investigation, in an article in the *Revue des deux mondes*, August 1920.

On the Reds' side the basic material is an article included in a work by P. Bykov, *Rabochaia revoliutsia na Urale* ('The Workers' Revolution in the Urals'), published in Ekaterinburg itself in 1922, and reissued in an expanded version in 1926, under the title *Poslednie dni Romanovykh* ('The Last Days of the Romanovs').[3] Bykov was one of the members of the Ekaterinburg soviet that ordered the execution in the summer of 1918.

These two documents, Sokolov's and Bykov's, have points in common, and subsequently the accepted version of the event was constructed on the basis they provided.

2. *Enquête judiciaire sur l'assassinat de la famille impériale russe*, Paris, 1924, 324 pp.
3. Sverdlovsk, 136 pp.

Meanwhile, the appearance of a young woman who claimed to be Anastasia, the Tsar's daughter, 'the only one who escaped from the massacre', made newsworthy these unburied dead. However, the most respected experts declared that the conditions in which the execution was carried out were such that nobody could have escaped. Convicted of imposture, but refusing to admit it, 'Anastasia' passed from the 'political' section of the newspapers to the column of 'news in brief'.

That was still the situation in 1975 when two BBC journalists, Anthony Summers and Tom Mangold, managed to get hold of the complete file of the investigation, the material from which Sokolov had composed his book. By studying the data they were able to show that the *Inquiry* published in 1924 had systematically eliminated those documents that might support the thesis that the Tsar's wife and daughters had survived. What had been forgotten was that the first communiqué announcing the execution had added: 'His wife and son have been sent to a place of safety.'

What is the value of these allegations? We are now in a better position to form a judgement in that almost the entire file of the investigation has been published by Nikolai Ross, in Germany, under the title *Gibel' Tsarskoy Sem'i* ('The Destruction of the Imperial Family').[4] In terms of volume, the report published by Sokolov is only about one-tenth of the file published by Ross, which shows the importance of the latter, very carefully produced publication. These documents, like the earlier ones, furnish the elements for an analysis of the immediate data regarding the assassination, since they are evidence turned up by the investigation; but in themselves they do not enable us to set the problem in its political context. To do that, other testimonies and records must be adduced.

Let us look first, however, at the information we possess concerning the end of the Romanovs and its circumstances.

'During the night of 16–17 July, at Ekaterinburg, by decision of the Presidium of the regional soviet of workers', peasants' and Red

4. Frankfurt, 1987.

Army men's deputies for the Ural region, the former Tsar Nicholas Romanov II was shot. This crowned murderer had lived too long, thanks to the kindness of the revolution.' On page 3 of the same issue of the newspaper *Uralsky Rabochy* for 23 July 1918 the following appears under the heading 'Telegrams':

Chairman Sverdlov announced receipt by direct wire of advice from Ural Regional Soviet of shooting of ex-Tsar Nicholas Romanov. In recent days danger of approach by Czechoslovak bands posed serious threat to Ekaterinburg, capital of Red Urals. At same time, fresh plot by counter-revolutionaries exposed, aimed at tearing crowned executioner from hands of soviet power. In view of these circumstances presidium of Ural Soviet decided shoot Nicholas Romanov. Execution carried out 16 July. Wife and son of Nicholas sent to safe place. Documents concerning plot sent to Moscow by special courier ... It had recently been proposed put former Tsar on trial for crimes against the people. However, current events prevented trial being held. Presidium having discussed reasons that led Ural Soviet to take decision to shoot Romanov, Central Executive Committee, through its presidium, decided accept decision of Ural Regional Soviet as correct.

In Moscow, the *Izvestiya* of 20 July had accompanied this information with the commentary: 'By this act of revolutionary punishment, Soviet Russia has given a solemn warning to all its foes who dream of restoring the former Tsardom, or who even dare to attack us.'

This first document thus associates the execution with the danger of a plot or an attempt to rescue the Tsar – perhaps another attempt similar to the Yakovlev affair? It mentions the approach of the Czechoslovak troops, more or less linked with the Whites, and the possibility that they might carry off the Tsar. It is composed in such a form as to show that the initiative and procedure for the execution came from the local authorities: 'their decision was correct'. All these points need to be checked.

In 1932 the British agent Bruce Lockhart stated that he was the first to be told the news, by Karakhan, Chicherin's deputy as People's Commissar for Foreign Affairs:

On the evening of July 17th I received from Karakhan the official intimation of the murder of the Tsar and his family at Ekaterinburg. I believe that I was the first person to convey the news to the outside world. The only first-hand evidence of this crime that I can give concerns the official attitude of the Bolshevik government. My own impression is that, alarmed by the approach of the Czech troops, who ... were now at open war with the Bolsheviks, the local soviet had taken the law into its own hands, and that the approval of the Central Government was subsequent. Certainly, there was no question of disapproval or disavowal ... Karakhan, it is true, professed to be shocked and pleaded extenuating circumstances.

These circumstances were indeed dramatic, both for the Soviet government and, still more, for the Bolsheviks of the Ural district. In all their history the Bolshevik rulers of Russia were never in such danger as at that time, not even during Admiral Kolchak's victories in 1919.

In June and July of 1918 the Bolsheviks controlled only a territory roughly equivalent to medieval Muscovy. Since April 1918 the Ukraine had been in the hands of Hetman Skoropadsky, who was in turn dominated by the Germans. A number of Russian monarchists, including Milyukov, the former Minister of Foreign Affairs, had taken refuge in Kiev, where they were negotiating with the Germans, but also with the volunteer army of Generals Alexeyev and Denikin and with the Don Cossack Ataman Krasnov, who held all south Russia as far as the Caspian Sea. Another front of civil war had been opened in western Siberia, where members of the former Constituent Assembly dissolved *manu militari* by Lenin found themselves associated, willy-nilly, with counter-revolutionaries and were trying to form an underground army. An unexpected event helped them – the revolt of the Czechoslovak soldiers, former Austro-Hungarian prisoners of war, who had been given permission by Trotsky to make their way to Vladivostok. Faced with the general disorder and with the ill will and mistrust of the more or less Bolshevik local soviets, they had seized Omsk with the help of the anti-Bolshevik railway workers. Gradually, this Czechoslovak force, under the command of General Gajda and others, became the only properly organized army in the western part of Siberia.

At the end of June and the beginning of July 1918, these soldiers, supported from Vladivostok by the Allies, increased their power, occupying one after another the stations along the Trans-Siberian Railway. They advanced from Omsk towards Ekaterinburg, where the imperial family were living.

At the same time, scattered clandestine groups of SRs were carrying on a struggle against the Bolshevik 'usurpers'. They acted as they had acted before 1914, by means of punitive raids and mass strikes, as at Izhevsk, with the support of democrats and of petty bourgeois victims of Red Guard excesses. Active among them was Boris Savinkov, the terrorist of 1900–05 who later became one of Kerensky's ministers and then an ally of Kornilov. He founded the Union for Defence of the Fatherland and Freedom. More or less associated with General Denikin, his network covered almost the whole of Russia and was supported by the French intelligence service, who, operating from Archangel, were expected to help Savinkov to march on Moscow. At the critical moment, however, this French aid failed to materialize, and Savinkov seized control of Yaroslavl, in a surprise coup, all on his own. The operation failed. The festival organized in his honour and the consternation that followed the recapture of the town by the Bolsheviks revealed how unpopular their regime was in the smaller provincial towns.

The map shows us that Yaroslavl, Izhevsk and Omsk are all in the same general area as Ekaterinburg, where the imperial family were imprisoned. All these events took place in June or early July, Yaroslavl having been seized by Savinkov on 6 July 1918. On that same day the Left SRs, who had supported the Bolsheviks in October and when the Constituent Assembly was dissolved, but had broken with them over the treaty of Brest-Litovsk – which they saw as a stab in the back for the German proletariat, since it strengthened the Kaiser's position – organized a successful attempt on the life of Mirbach, the German Ambassador in Moscow, followed by an attack on the HQ of the Cheka. Their aim was obvious – to launch a revolutionary war against the Germans, who were re-establishing the old regime in the territories they occupied.

All these circumstances, taken together, explain the measures

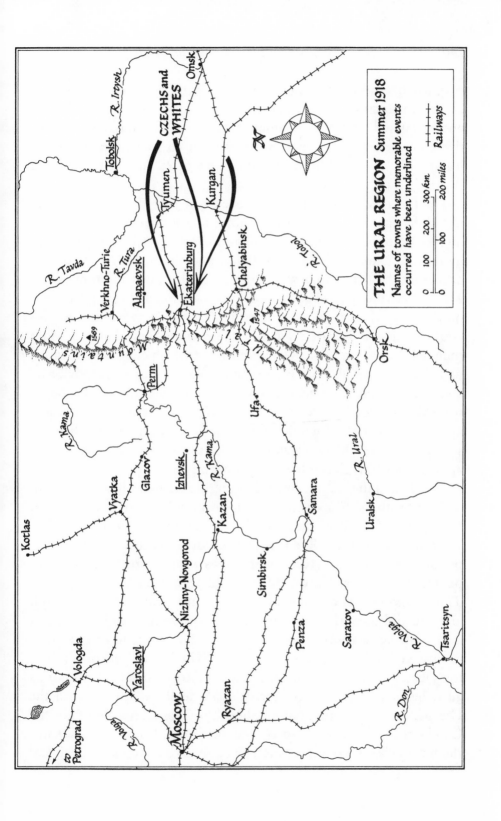

adopted by the Ekaterinburg soviet, directly threatened as it was by the advancing Czechs. The soviet executed the Tsar on 16 July and evacuated the city in the immediately following days. The executive committee of the soviet were the last to leave, like a captain leaving his sinking ship, during the night of 25 July. They went to Perm.

The facts we possess about the events of July 1918 are doubtless not sufficient, by themselves, to account for the execution of the imperial family, but they do explain at least two features of what happened, namely, the improvisation and the tempo. The idea of executing the Romanovs was certainly in the air at the time. On 4 March 1918 the Bolsheviks of Kolomna had already 'demanded' this, on the grounds that 'the German and Russian bourgeoisies are re-establishing the Tsarist regime in the regions they occupy'. More than once, a trial in Moscow, the verdict of which one could imagine, had been proposed. The Ekaterinburg authorities were obviously aware of all that. The chairman of the local soviet, A. Beloborodov, was close to Trotsky, who, it was said, would act as public prosecutor at the Tsar's trial. It would seem that circumstances impelled the soviet to act, and Moscow endorsed the action *post-factum*. All this appears probable, but it has not yet been proved.

Hardly had the Whites entered Ekaterinburg than they discovered, in Four Brothers' Wood, burnt vestiges of clothes and other personal belongings of the Romanovs. An investigation was set on foot.

At the time, while in Red Russia, according to Bruce Lockhart, 'the population of Moscow received the news with amazing indifference', people in the West learned of it on 22 July from a report in *The Times*: 'Ex-Tsar Shot. Official Approval of Crime'. This paragraph and the obituary that accompanied it occupied a column and a half. Nothing was said about the family, the report having referred exclusively to the execution of the Tsar. The fate of Nicholas II was the subject of comment on page 5. The editorial in *The Times* of 26 July, headed 'Prompt Help for Russia', dealt exclusively with the Czechs, and said nothing at all about the Tsar's fate. Were not the Czechs reconstituting a sort of 'second front', behind the backs of the Bolsheviks, whom the Allies, since Brest-Litovsk, saw as virtual accomplices of Germany? The way the war was going certainly interested the Allies more at this time than the personal destiny of the Romanovs. The second battle of the Marne, in which the outcome of the conflict was again to be decided, was going on at this very moment.

However, the question of what had happened to the Romanovs came up as soon as the war was over.

Immediately after the Armistice, at the tribune of the Chamber of Deputies in Paris, the Minister of Foreign Affairs, Stephen Pichon, gave the first public account of the murder of the Romanovs, which he had obtained from Prince Lvov, who had been the first head of the Provisional Government, in March 1917.

Prince Lvov was in a cell adjoining the one where the members of the imperial family were held. The Bolsheviks brought the family together, made them sit down and then, during the night, bayoneted them, after which they finished them off, one by one, with revolver shots. The room, I was told by Prince Lvov, was a veritable pool of blood.

This public statement naturally had considerable effect, as the account given by a minister who was quoting the first head of Russia's Provisional Government.

Actually – though this did not become known till much later – Prince Lvov had never lived in the Ipatiev house, where the imperial family were imprisoned. He had never even entered it. Moreover, it did not contain any 'cells', being an ordinary bourgeois residence. Pichon must have misunderstood. Lvov had certainly been in a cell, but this was four kilometres distant from the Ipatiev house, and he had not witnessed the events he described. However, he would not retract his statement.

In any case, the Bolshevik leaders denied that they had killed the whole family – Chicherin in the *New York Times* of 20 September 1918, then Maxim Litvinov in a statement made on 17 December 1918. The most explicit statement from that quarter was, however, the interview given by Chicherin to the *Chicago Tribune* at the

Genoa conference, which was reproduced in *The Times* of 25 April 1922:

Question: Did the Soviet Government order or authorize the killing of the young daughters of the Tsar, and, if not, have the perpetrators been punished?

   Reply: The fate of the young daughters of the Tsar is at present unknown to me. I have read in the Press that they are now in America. The Tsar was executed by a local soviet without the previous knowledge of the Central Government. The event took place on the eve of the occupation by the Czechoslovaks of the whole region. A conspiracy had just been discovered to free the Tsar and his family and bring them over to the Czechoslovaks ... At a later date, after having become acquainted with the essential facts bearing on the case, the Central Executive Committee approved the execution of the Tsar. No reference was made to his daughters. Communication with Moscow having been stopped immediately after the execution of the Tsar owing to the Czechoslovak occupation, the circumstances of the case have not yet been cleared up.

Were these statements intended for the foreign press? The first was made while the war was still on. Chicherin's interview at Genoa was subsequent to the 'reappearance' of the alleged Anastasia in Germany.

   In any case, General Dieterichs denied that the Bolsheviks were unaware of the fate of the Romanov women. He spoke as one well informed, since it was he who appointed N. Sokolov to investigate the affair. His testimony was included in an article entitled 'Le Crime d'Ekaterinbourg', by Nicolas de Berg-Poggenpohl, in the *Revue des deux mondes* for 1 August 1920.

The Bolsheviks announced that the Tsar was dead, but denied that this was true also of the other members of the Imperial family and their attendants. They did everything they could to mislead public opinion. For example, on 20 July, three days after the crime, a train left Ekaterinburg and it was announced, with great publicity, that this train was carrying the imperial prisoners. Actually, the only passengers in this train, which set out for Perm, were the Tsaritsa's reader and friend Mademoiselle

Schneider, the young lady in waiting Countess Hendrikov, the butler Nagorny, and the footmen Volkov and Trupp. All of them, except for one of the servants who managed to escape, were shot near Perm on 22 August 1918.

Nicolas de Berg-Poggenpohl, who 'presented' this statement, added the following comment:

It is to be hoped that this denial will kill once for all the rumours and fables that continually reappear – and always from a Bolshevik source – according to which the Tsar is still alive. One of those articles was published in Moscow on 17 December 1918: Litvinov (Finkelstein), in Copenhagen, admits one part of the murder and denies the other. In a German newspaper in April 1920 there appeared a letter from an alleged former prisoner of war who claimed to have been present at the murder of Nicholas II alone . . .

The reason for these tendentious rumours is clear to anyone who knows the Russian soul – to create the maximum confusion, dissension, fear and superstitious hope in these minds that have already been so deeply shaken and affected to their very roots.

Dieterichs's testimony has been largely contradicted by documents published since. Thus, A. Volkov was alive on 23 August 1919, when he was interrogated by Sokolov in Omsk, and his testimony recorded (cf. Ross, document no. 256) – unless it was invented. K. Nagorny was certainly shot, but in May or at the beginning of June 1918, and so before the execution of the Romanovs (cf. Ross, document no. 15, p. 63).

Countess Radziwill replied to an American journalist who asked her to tell about events in Siberia in summer 1918, and showed her the first photos of the discoveries made in Four Brothers' Wood, that the jewellery found there was 'not in the possession of the Empress Alexandra when the imperial family was arrested and therefore she could not have taken it with her to Siberia'. The Soviet authorities had seized those jewels and must have left them where they were found in order to create a false impression. She added that Princess

Dolgorukov, whose husband was reported to have been killed with the imperial family, had recently received a message from him, and the relatives of Countess Hendrikov, also allegedly killed, had learned she was still alive. Countess Radziwill further said that she had heard that Zinoviev had told 'a foreigner who happened to be in Petrograd a short time after the alleged murders' that only the Tsar had been executed; his family 'were living in safety in a city in Siberia, the location of which he refused to disclose'.

The American journal (*San Francisco Sunday Chronicle*, 11 July 1920) in which this interview appeared asked the question 'Was the Tsar's family really slain? Proofs of death ineffective'.

Rumours then began to circulate saying that the *entire* imperial family had been able to escape. In a book published in 1920 Major Lasies, of the French military mission in Siberia, wrote:

On 12 May 1919 I left Ekaterinburg with General Janin and went to the headquarters of General Pepelaev ... In the course of this day I had occasion to talk about my visit of the previous day [to Ekaterinburg] with Lieutenant X, who spoke French very well ... 'You are doubtful about the death of the imperial family?' he said. 'Perhaps you are right ... ' He then read to me two letters from a member of his family.

A letter dated April 1919 contained the following: 'The Emperor is here! How is one to understand this? I think you will understand it in the same way as we have understood it. If the fact is confirmed, the Easter festival will be bright for us, and infinitely joyous.' The other letter was dated a few days later, and said: 'In the last few days we have had confirmation regarding the health of those whom we love. God be praised!'

Lasies goes on:

My friend Major Bolifraud, who represented the French military mission at Ekaterinburg over a long period, and who visited the Ipatiev house with me, wrote me the following letter ...

'Paris, 24 March 1920. Dear Friend. You ask me to recall our conversation on the platform of Ekaterinburg station in May 1919 ... I believed in good faith in the story as it was told to me, even though I could never obtain any first-hand evidence. I began to feel some doubt only when I read the official report by the examining magistrate.

In the end, Lasies tells us, after a second visit to Ekaterinburg, he came to 'believe in the Tsar's death'. Were not these words written in German on a wall in the Ipatiev house: 'That very night the Emperor was shot.'[5] 'If the imperial family had suffered the same fate,' he remarks, 'the inscription would doubtless have mentioned the fact ... I am quite ready to believe that the murderers managed to make one corpse disappear, but it is inconceivable that they could have done the same with thirteen!' And he is sceptical that they could have buried, disinterred and reburied these corpses in several different places, as alleged by the witness Kukhtenkov. In any case, the investigating magistrate told him that the bodies 'had been burned'.

This magistrate said at the end of his report that one of the actual perpetrators of the crime would soon be questioned. ' "So, then," I said to him, "here you have at last the first-hand witness. What emerged from his revelations? What did he say?" "Alas," the magistrate replied, "he died of typhus before he could reveal anything!" '

Were they all still alive? Or were they all dead? Burned? Or beheaded, as asserted by Essad Bey, who reports this story by the monk Iliodor:

'I had to go to the Kremlin to see Kalinin [chairman of the Central Executive Committee of the Soviets] and talk to him about some important church reforms. Passing through a dark corridor, my guide suddenly

5. The inscription on the wall ran: 'Belsatzar ward in selbiger Nacht / Von seinen Knechten umgebracht/' ('That same night Belshazzar was killed by his slaves'). In the original verse by Heine the name is 'Belsazar'. The inscription shows, at least, that its author was someone of German culture, presumably a German prisoner of war, or perhaps a Latvian. It refers to the killing of the Tsar, and of the Tsar only. Does this mean that the writer knew that the execution had happened or was going to happen? Does the inscription imply that the executioners did not try to conceal their deed? Does it prove that the Tsar was executed in that same place?

opened the door to a small secret chamber. I entered. On a table, under glass, lay Nicholas II's severed head, a deep wound over the left eye. I was petrified.'

... According to rumours [says Essad Bey], the severed head was brought to Moscow at the order of the Ural soviet by the prostitute Gusseva, paramour of one of the alleged murderers. The journey with the head of the Anointed One proved too much for the woman. She lost her mind. Barefoot, her clothes in tatters, her hair flying wildly, she strode through the deep snow of Moscow and, in a babbling voice, told people congregating around her that she had brought back the head of the Anointed One to the holy city of his coronation. Eventually she was shot and her story perished with her.

This story, which seems taken from a chronicle of the time of Ivan the Terrible, includes elements drawn from familiar legends: the severed head, the dark corridor, the prostitute, the wanderings about, the madness. It is nevertheless corroborated by two testimonies, one from General Dieterichs and the other from his friend Robert Wilton, the correspondent of *The Times*, who wrote that Yurovsky 'left [Ekaterinburg] on the 19th [July 1918] with the plunder and, it is believed, the "heads"'.

In 1924 the conclusions of the magistrate N. Sokolov on the murder of the imperial family at last appeared in print. The book mentions the name of the organizer of the killings. 'The man entrusted with organizing the massacre was a Jew, Jacob Yurovsky, who had been a photographer and then a nurse and was a member of the Ural soviet. The workers originally assigned for the task were replaced by Cheka personnel, mostly Latvians.'

The story of the execution is related by Sokolov on the basis of the testimony of P. Medvedev, a welder and 'sole eye witness'. We can compare this with the version put out by the Bolsheviks.

In the evening of 16 July, between 7 and 8 p.m., when the time for me to go on duty had just come, Commandant Yurovsky ordered me to take all the Nagant revolvers from the guards and to bring them to him. I took twelve revolvers from the sentries and from some others of the guards, and brought them to the commandant's office. Yurovsky said to me: 'We must shoot them all tonight, so notify the guards not to be alarmed if they hear shots.' I understood, therefore, that Yurovsky had it in his mind to shoot the whole of the Tsar's family, as well as the doctor and the servants who lived with them, but I did not ask him where or by whom the decision had been made. I must tell you that, in obedience to Yurovsky's order, the boy who assisted the cook was transferred in the morning to the guard room [in the Popov house]. The lower floor of Ipatiev's house was occupied by the Latvians from the Latvians' commune who had taken up their quarters there after Yurovsky was made commandant. They were ten in number. I don't know the names of any of them.

At about ten o'clock in the evening, in accordance with Yurovsky's order, I warned the guards not to be alarmed if they should hear firing.

About midnight, Yurovsky woke up the Tsar's family ... In about an hour the whole of the family, the doctor, the maid and the two waiters got up, washed and dressed themselves. Just before Yurovsky went to awaken the family, two members of the Cheka arrived at Ipatiev's house. One of them I learned later was called Pyotr Yermakov, but I don't know the other's name ...

Shortly after one o'clock the prisoners left their rooms. The Tsar carried the heir in his arms. They were both dressed in soldiers' shirts and wore caps. The Empress and her daughters wore neither coats nor hats. The Emperor, carrying his heir, preceded them. The Empress, her daughters and the others followed him down. Yurovsky, his assistant and the two Chekists accompanied them. I was also present.

... Having descended the stairs ... we entered the ground floor of the house. Yurovsky led the way into the room that adjoins the lumber-room and ordered chairs to be brought. His assistant brought three, which were given to the Tsar, the Empress and Alexei. The Empress sat by the wall, where there is a window, near the back pillar of the arch. Behind her stood three of her daughters (I knew their faces very well because I had seen them nearly every day when they walked in the garden, but I didn't know their names). The Emperor and his son sat side by side almost in the middle of the room. Dr Botkin stood behind Alexei. The maid, a very tall woman, I don't know her name, stood at the left of the door leading to the lumber-room, with, by her side, the fourth of the Grand Duchesses. The two servants stood in the left-hand corner facing the entrance, near the wall separating the room from the lumber-room.

The maid carried a pillow. The Tsar's daughters had also brought small cushions with them. They put one on the Empress's chair and another on the chair where Alexei was sitting ...

At this moment eleven men entered the room: Yurovsky, his assistant, the two Chekists and seven Latvians. Yurovsky ordered me: 'Go into the street, see if there is anybody there, and wait and check whether the shots are heard.' I went out into the yard ... and before I got to the street I heard firing. I turned back immediately (only two or three minutes had elapsed) and ... saw that all the members of the Tsar's family were lying on the floor with many wounds in their bodies. Blood was running in streams. The doctor, the maid and the two servants had also been shot. When I entered the heir was still alive and moaning a little. Yurovsky went up and fired two or three more times at him, point blank. Alexei then lay still.

The sight of the murder and the smell of blood made me feel sick. Before the killing, when Yurovsky distributed the revolvers, he gave me one, but, as I said before, I took no part in the murder. Besides his Nagant, Yurovsky had a Mauser. After the assassination Yurovsky told me to bring some guards to wash the floor. On the way to Popov's house I met two of the senior guards, Ivan Starkov and Konstantin Dobrynin. They were running in the direction of Ipatiev's house. Dobrynin asked me: 'Has Nicholas II been shot?', adding: 'Take care that they haven't shot somebody else instead. You are responsible.' I declared that the Tsar and all his family had been shot.

I brought twelve or fifteen guards back with me to the house. These men carried the dead bodies out to the lorry that was waiting near the entrance to the house, on stretchers made from bedsheets and shafts of sledges taken from the yard. They were wrapped in cloth for soldiers' uniforms which was found in the lumber-room. The driver's name was Lukhanov, he was a workman from the Zlokazov factory. Yermakov and the other Chekist got into the lorry and it drove off. I don't know where they went or what was done with the bodies.

The blood in the room and the yard was washed away and everything was put in order. By three o'clock in the morning the job was finished. Yurovsky went to his office and I went back to the guard room.

I woke at nine and went to the commandant's office. Present there were the chairman of the district soviet, Beloborodov, Commissar Goloshchekin and Ivan Starkov, the guard commander ... All the rooms in the house were in great disorder. Things were scattered about everywhere. Suitcases and trunks had been opened, and piles of gold and silver articles lay on all the tables ... It did not interest me who had decided the fate of the imperial family, and by what right. I just carried out the orders of those in whose service I was.

Among the Bolshevik leaders, Beloborodov and Goloshchekin often came to the Ipatiev house.

This very detailed account raised the question: did Medvedev take an active part in the murder of the imperial family or was he, as he claims, a mere spectator? When he left Ekaterinburg with the Reds, Medvedev abandoned his family at Sysert. His wife Maria was interrogated on 9 November 1918. This was her deposition:

The last time I came to my husband in the city was in the early part of July of this year (reckoning by Old Style) ... Left alone with me, my husband told me that the Tsar, the Tsaritsa, the Tsarevich, all of the Grand Duchesses and all the servants of the imperial family had been killed several days before. My husband did not give me the details of the murder at that time. In the evening my husband sent the detachment to the railroad station. On the following day we went home because the commandant had given him two days' leave so that he could distribute money to the families of the Red Army personnel.

Once home, Pavel Medvedev told me several details of how the murder of the Tsar and his family was accomplished.

According to Pavel, during the night, at about two o'clock, he was ordered to awaken the Tsar, the Tsaritsa, all the imperial children, those with them and their servants ... All those who were awakened got up, washed, dressed and were taken to the lower floor, where they were placed in one room. Here they were read a paper which said: 'The revolution is dying, and you must die too.'

After that they began to shoot at them and they were all killed. My husband also fired. He said that he was the only one of the Sysert workers who took part in the shooting ...

They took the murder victims a long way into the woods and threw them into some holes ...

My husband told me all this quite calmly. Towards the last he became unresponsive, did not recognize anyone and acted as though he had ceased to have any concern for his family.

This testimony was recorded three months before that of Pavel Medvedev. It is to be observed that, contrary to her husband's story, this witness says that he took part in the shooting. Also, Medvedev claimed not to know what had been done with the bodies. The version according to which they were buried is supported by the testimony of P. Kukhtenkov, who deposed on 13 November 1918 before Judge Sergeyev:

I had been working for ten years at the Verkh-Isetsk factory ... when the October Revolution took place. Engineer Dunayev was put in charge, and the Factory Committee appointed me to be commissar, to keep an eye on him ... But I wasn't any good at that and, six weeks later, I joined the Red Army, having become a Party member in January 1918. At the front I was sent to fight the army of General Dutov, at Troitsk. Then I was sick and in hospital. After that I was discharged from the army ... I was given a job at the Party Club in Ekaterinburg, which was frequented also by non-members of the Party ... Among the men often there were the members of the Presidium: Paramonovich (or Parmenovich), Sibrin, I. Frolov, A. Kostousov and others whose names I don't know. They sometimes stayed overnight, when a guard was mounted. On 18 or 19 July, about 4 a.m., the chairman of the Verkh-Isetsk soviet, Malyshkin, arrived, together with the military commissar Yermakov and other prominent Party members ... When I passed them on my way to put out the lights, one of them said: 'Leave us, Comrade Kukhtenkov, we have to talk business' ... I went out to do some shopping in the market, because I lived nearby ... When I got back they were gone.

Next day, at 4 a.m., the same men, except Malyshkin, came to the club. They had a warlike air about them. This struck me and I wanted to know what was up ... It was already light, and I went close to them to put the electric light out ... I heard someone say: 'Altogether, that makes thirteen.'

When they noticed me, one said: 'Leave us ... No, we'll go into the garden. You tidy the place up.' I pretended to tidy, then went into the bathroom, and on through the hen-house into the little garden that is behind the one where the men were talking together ... From there I could hear everything they said. A. Kostousov said: 'Another day of messing about. Yesterday we buried, today we rebury' ... I heard only a few sentences. I realized that they had taken part in the burial, because I heard: 'The bodies were still warm ... I felt the Tsaritsa's and it was still warm ...' Then: 'They were all wearing drawers.' One said that the Tsarevich had died at Tobolsk ... Another said that there were Nicholas, Sashka [the Tsaritsa], Tatiana, the Heir and Vyrubova ...

A few days after that the Czechs arrived ... I had left the town to fetch milk and was arrested.

(Ross, document no. 65)

What impresses in this testimony is, first, the curriculum vitae of this worker, typical of the year 1918 when workers were indeed appointed to supervising jobs which were beyond their capacity. But, as regards the facts narrated, we remain puzzled because the names he mentions are not met with elsewhere, apart from that of Yermakov, who, according to another testimony, was executed in Odessa. Medvedev, too, died before Judge Sokolov could hear what he had to say.

Let us now compare these testimonies with the account given by members of the executive committee of the Ekaterinburg soviet, through Bykov. This version ignores completely the statements made by Zinoviev, Chicherin and Litvinov. It states that *all* the Romanovs were executed and, in the main, coincides with Sokolov's version, though it diverges on some significant points. It dates from 1921 and is thus contemporary with Sokolov's report, at any rate before that report was made public.

The question of shooting the Romanovs was raised at meetings of the regional soviet as early as the end of June. The SR members – Khotinsky, Sakovich ... and others, ... insisted on immediate execution ... The question of shooting Nicholas Romanov and all who were with him was decided in principle in the first days of July. The presidium of the soviet was entrusted with the task of organizing the execution and determining the day for it. The sentence was carried out during the night of 16–17 July. At the meeting of the presidium of the All-Russia Central Executive Committee on 18 July the chairman, Ya. A. Sverdlov, announced that the ex-Tsar had been shot. After considering all the circumstances that had compelled the Ural regional soviet to take the decision to shoot Romanov, the presidium of the All-Russia Central Executive Committee gave its approval to this decision.

The organization of the execution and the destruction of the bodies of those shot were entrusted to a reliable revolutionary who had been in action at the front against Dutov, a worker at the Verkh-Isetsk factory named Petr Zakharovich Yermakov. The actual execution of the ex-Tsar had to be carried out in conditions such that no active intervention by supporters of the Tsarist regime would be possible. That was why this method was chosen.

The Romanov family were told that they had to come down from the upper floor, where their rooms were, to the ground floor. The entire household came down at about 10 p.m. – the ex-Tsar, his wife, his son, his daughters, Botkin the family doctor, the Tsarevich's attendant and the ex-princesses' maid. They were all in their usual clothes, as they always went to bed later than that.

There, in one of the rooms in the semi-basement, they were put against a wall. The house commandant ... read the death sentence and added that the Romanovs' hopes of liberation were vain, that they were all to die.

This unexpected information left them stunned and only the ex-Tsar managed to say, questioningly: 'So we are not being taken anywhere?'

The condemned persons were killed with revolver shots. Only four men were present at the execution, and all of them fired.

At about 1 a.m. the bodies were transported out of the town into the woods, in the neighbourhood of the Verkh-Isetsk factory and Palkina village, where they were burned on the following day.

The shooting itself had not been heard by anyone, even though it took place almost in the centre of the town. The noise had been covered by that made by the engine of a car stationed under the windows of the house during the execution. Even the men guarding the house remained unaware of what had happened, and two days later they were still taking over from each other at their posts outside the house.

Later, the attempt by the White military authorities to find the bodies 'was without result'. 'The Romanovs had been shot wearing their ordinary clothes. When it was decided to burn the bodies, they were first stripped. In some parts of the clothing, which was subsequently burned, jewels were found to have been sewn. It may be that some of these dropped out or were put on the bonfire with the clothes.'

This account by Bykov goes on to say that Michael Alexandrovich, the Tsar's brother, was also shot, and he mentions the slaughter of the Grand Dukes at Alapayevsk.

It must be recorded that official Soviet announcements did not publish full information about the execution of the Romanov family at the time when this occurred. Mention was made only of the execution of the ex-Tsar: the Grand Dukes were said either to have escaped or to have been carried off by persons unknown. The same was said about Nicholas's wife, son and daughters, who were supposed to have been taken away 'to a place of safety'.

This was not due to lack of resolution on the part of the local soviets. The historical facts tell us that our soviets – the regional soviet and the soviets of Perm and Alapayevsk – acted boldly and decisively, having resolved to destroy everyone who was close to the throne of the Autocracy.

Between the White version and the Bolshevik version there are more similarities than differences. In both, the entire family is murdered, it happens during the night, and subsequently the remains are taken to a place some distance away. In both versions Yermakov plays a central role, at least as an agent; but whereas in the White version he is helped by 'the Jew Yurovsky', the latter is absent from the Red one. Also, the account published on the spot speaks of four executioners, and not of eleven, as in Medvedev's story. It mentions particularly the influence of the Left SRs in the development of the events, and does not mention A. Beloborodov, the chairman of the presidium.

The edition (somewhat enlarged) of the Bolshevik account that was published in 1926 mentions explicitly Leon Trotsky as the one who was to act as prosecutor at a trial that, it turned out, could not be held. F. Goloshchekin, who was a member both of the presidium of the Ural soviet and of the military committee, played a central role, it would seem, in that he went to Moscow in June 1918, to discuss, *inter alia*, the question of what was to be done with Nicholas II. He returned to Ekaterinburg on 14 July, two days before the execution, and was present on the 18th and 19th, during the incineration of the corpses. This document makes more dubious the hypothesis of a decision taken independently and in haste by the local authorities. It mentions also the appointment of Yurovsky and of G. Nikhin to the local Cheka, and puts some of the responsibility for what was done on the Left SRs and the anarchists. 'Not being convinced that the Bolsheviks would shoot the ex-Tsar, they decided to act on their own and with their own means.' But neither the White commandos who wanted to rescue the Romanovs nor the Left SRs and anarchists who wanted to kill them were able to do what they wanted; and, in the end, 'after studying the military situation, the soviet took the decision not to await the trial but to execute Nicholas II'. It is to be noted that the extract from the newspaper *Uralsky Rabochy* of 23 July 1918 published in this pamphlet speaks only of the execution of Nicholas II, whereas the document itself makes explicit the execution of the entire family.

Some problems arise from these documents published at Ekaterinburg-Sverdlovsk. The first is the use made, in the 1926 publication, of the book by the White General Dieterichs, which is quoted several times as a source of information. Then there is the total absence of any reference to the statements made by Zinoviev, Litvinov and Chicherin concerning the survival of the wife and daughter of Nicholas II. The 1921 edition had indicated that it had been announced that some members of the family were 'taken to a place of safety', but that this was done so as not to reveal immediately that, in fact, they had all been killed. Sokolov supported this thesis on the basis of another telegram that allegedly informed Moscow 'that the princesses had been killed while attempting to escape'. However, as Summers and Mangold have shown, this telegram was later exposed as a forgery.

The emphasis on the role of the Left SRs is surprising, given that the Bolsheviks eventually did what they wanted done. Was it Bykov's intention to show that the Left SRs failed to take over the initiative from the Bolsheviks and that it was the Bolsheviks, and they alone, who decided and acted? Besides, it is not clear why, the execution of the Tsar having been made public, together with the executions at Alapayevsk of Grand Dukes Sergei Mikhailovich, Constantine Konstantinovich, Igor Konstantinovich and Vladimir Pavlovich Paley, along with Grand Duke Sergei Alexandrovich's widow, Grand Duchess Elizabeth Fyodorovna, the Tsaritsa's sister (officially, by rebel bands) and also the execution of the Tsar's brother Michael, the fate of Alexandra and her daughters should never have been mentioned – except to say that they were safe or, in the (forged) telegram, that all the family were dead.

The precise role played in this affair by the central government gives rise to another question. The Bolshevik documents of 1921 and 1926, originating at Ekaterinburg, allot a share to local initiative and a share to instructions received from Moscow, but without revealing what those instructions said.

We know that there had been talk of a trial, that Trotsky had proposed that this trial be held in public and that the proceedings would have been broadcast through loudspeakers all over Russia (doubtless through recordings). But Lenin considered that in the summer of 1918 there were other priorities. Trotsky recalled in 1935 that, after the loss of Ekaterinburg, he asked Sverdlov:

'Oh yes, and where is the Tsar?' 'It's all over,' he answered; 'he has been shot.' 'And where is the family?' 'And the family along with him.' 'All of them?' I asked, apparently with a touch of surprise. 'All of them!' replied Sverdlov: 'What about it?' He was waiting to see my reaction. I made no reply. 'And who took the decision?' I asked. 'We decided it here. Ilyich believed that we shouldn't leave the Whites a live banner to rally around, especially in the present difficult circumstances.'

Did that mean that the Ekaterinburg Bolsheviks were directly ordered to execute the Tsar, or merely that they were ordered to execute him if circumstances made this necesary? 'You were right to do what you did,' said Sverdlov's telegram to the Ekaterinburg soviet. If anyone was in a position to know the truth of the matter, it was Sverdlov. But did he tell Trotsky the truth? Or is it Chicherin, according to whom the Empress and the daughters had *not* been executed, who should be believed?

Lenin referred publicly to the Tsar's death in his address to poor peasants' delegates on 8 November 1918. Before that, in July and August, he accused the Left SRs of wanting to restore Tsarism, acting as allies of the Czechoslovaks and the Anglo-French imperialists. He added that he might make an alliance with the Germans ('Letter to American Workers', 22 August 1918): a startling formulation.

One hypothesis would be that the Ekaterinburg Bolsheviks did indeed act, but precisely in order to prevent what the Left SRs wanted to do on their own – that is, they executed the Tsar, Alexei and some of the Tsar's staff, but saved the women before the Left SRs could realize their objective of killing everyone. This hypothesis, which I shall try to test, would explain the Bolsheviks' concern to insist that the entire family had been executed. The announcement would not only have meant that 'there were no Romanovs left' but also would have served to put the Left SRs off their stroke; they were, in any case, arrested soon afterwards. As for Khotinsky and Sakovich, Left SRs both, they were, according to the Reds, shot by the Whites. Let us recall the circumstances.

On 6 July the Left SRs had just murdered Mirbach, and the Bolsheviks could have reason to fear that at Ekaterinburg they might also execute Nicholas's 'German' wife and his daughters, all of whom might be seen as belonging to the House of Hesse. The Tsaritsa's brother, Ernst of Hesse ('Uncle Ernie') had another sister in Russia, the widow of Grand Duke Sergei Alexandrovich, who had been murdered in 1905. Another of Alexandra's sisters, Irene, was married to Wilhelm II's brother Henry Albert of Prussia, while the Kaiser was himself godfather to one of the Tsar's daughters. It is highly conceivable that Berlin considered it could not remain indifferent to the fate of 'the German women'.

It is also conceivable that their deaths, if admitted, might have endangered the peace of Brest-Litovsk; indeed, Mirbach had already said as much. On 4 July, Mirbach had, in his capacity as ambassador, attended the Congress of Soviets, where he was booed by the Left SRs. His murder was a catastrophe for Lenin and Sverdlov and they hastened to express their great regret to the authorities in Berlin. The members of the imperial family who were connected with the German Emperor must not suffer the same fate as Mirbach.

The 'unavowable' hypothesis is, thus, that the Bolsheviks wanted to execute the Tsar but to save the Tsaritsa and her daughters in order to keep in with the Germans. This could only be managed secretly, in a way that prevented the Left SRs from knowing what was happening, and with all the more care because the Yakovlev affair had already aroused much suspicion. This hypothesis, whether or not it is well founded, leaves unsolved the enigma of what became of the victims' bodies, to which I shall return. Commissar Voikov boasted that 'the world will never know what we did with them'. A curious remark, to be sure, but one that does not pre-judge the question of *how many* bodies disappeared.

For the Reds, a great deal would have depended on keeping secret what had been done, but for the Whites it would have been a matter of sacrilege. At the time, had the story been revealed, it would have confirmed all those allegations about the link between the Tsaritsa and the Germans. After the defeat of Wilhelm II, a family saved by the Kaiser and the Bolsheviks, those two 'enemies of the human race', would have been objects of opprobrium for the victors.

This hypothesis, unavowable for one side and sacrilegious for the other, has everything to arouse our curiosity. It is, of course, meaningful and open to testing only if two conditions are given: first, that we can establish what became of the survivors and, second, that we know how Nicholas died, since, in this case, he must have been executed otherwise than as we are told in the accepted version – either on his own, or along with Alexei and his staff.

The first person to advance this hypothesis was Captain Malinovsky, immediately after the murder, and on the basis of observations made on the spot.

Let us first recall what the situation was. On 25 July 1918 the Czechoslovak forces, along with the Russian contingents commanded by General Voitsekhovsky, entered Ekaterinburg. The Reds had already left the city, taking records and files with them to Perm. The executive committee of the Ural soviet and the local Cheka had also gone to Perm. A many-headed authority was set up in Ekaterinburg: the military government, a Czechoslovak national committee and a civil administration led by the former chairman of the Ekaterinburg stock exchange, P. Ivanov, a Cadet.

The new occupiers of the city knew that the imperial family had been there, and this had indeed speeded up the Whites' push towards Ekaterinburg. But they had come too late. They were told, and learned from the newspapers, that the Tsar had been shot and that his family had disappeared. The Ipatiev house, where they had been lodged, was empty. Neighbours and White officers were already visiting it in search of souvenirs. On the morning of the 27th a lieutenant reported to V. Girs, who commanded this sector, informing him that Reds had been seen at a point 18 kilometres from the city, near a spot known as Four Brothers' Wood, busy burning various objects. The garrison commander realized that these objects must be belongings of the imperial family, and wanted to inform the judiciary. In the absence, however, of an order from the public prosecutor, nobody would agree to take any action. The White officers, impatient, did not wait for this authorization but, on the responsibility of Captain Malinovsky, compelled the deputy public prosecutor, A. Nametkin, to draw up an inventory of what they had found near Four Brothers' Wood, to which they took him by force. The objects consisted of various articles and vestiges of clothing that had belonged to members of the imperial family.

Captain Malinovsky carried on investigating for six days and composed a report in which he expressed the view that a number of persons had been shot in the Ipatiev house in order to *simulate* the execution of the imperial family, that the family had been taken along the Koptyaki road and there undressed, and their clothing burnt, after which they were given peasant clothing and taken to some other place. 'These were my impressions as a result of my observation and considered thought,' he wrote. 'It seems to me that the German Imperial House could in no way permit such a wicked deed.' This document, which is in the Houghton Library at Harvard, has been published only by Summers and Mangold. It is not in Sokolov's report; and, oddly enough, in Ross's book the testimony of Malinovsky is cut short before this conclusion (see document no. 214).

The second disturbing fact is that Malinovsky's civilian assistant, Nametkin, who reacted to the discovery in the same way, was accused of cowardice and incompetence, and executed.

The third disturbing fact is that the first magistrate, Sergeyev, who was entrusted with the investigation by the public prosecutor, was soon removed from it by General Dieterichs, for lack of 'conviction'. This Sergeyev did not rule out the possibility that the whole family had been killed, but he was inclined to believe instead, after hearing several witnesses, Medvedev included, that the Tsaritsa and her daughters had not been executed but had been evacuated from Ekaterinburg to some other place.

Before being removed from the investigation, however, Sergeyev gave an interview to Herman Bernstein, of the *New York Tribune*, which appeared on 5 September 1920. He said:

I do not believe that all the ... people, the Tsar, his family and those with them, were shot there. It is my belief that the Empress, the Tsarevich and the Grand Duchesses were not shot in that house. I believe, however, that the Tsar, Professor Botkin, the family physician, two lackeys and the maid, Demidova, were shot in the Ipatiev house.

Sergeyev was himself shot, one month after giving this interview, on 23 January 1919. According to General Dieterichs, he was executed 'by the Bolsheviks'.

So, then, the suspect deaths already number five: V. Khotinsky and N. Sakovich, executed, according to the Reds, by the Whites;

P. Medvedev, dead from typhus between two spells of interrogation, according to Sokolov and Major Lasies; and A. Nametkin and I. Sergeyev, executed, according to the Reds, by the Whites.

Sergeyev's replacement in charge of the investigation was the magistrate Sokolov, who was convinced that all the Romanovs had been killed. This conviction was not shared by the military investigators who were the first to arrive on the scene in July 1918. Nor was it shared by the Whites' intelligence organization, which carried out its own inquiry at Perm as soon as this town had been occupied by General Pepelaev's forces in December 1918. The organization's chief, A. Kirsta, worked at the investigation from January to April 1919, but thereafter, on orders from General Dieterichs and Admiral Kolchak, Sokolov took exclusive charge of the inquiry into the fate of the Romanovs.

In the results of his inquiry that he published in 1924 Sokolov ignores the conclusions arrived at by the investigators working under Kirsta, who had found traces of the presence in Perm of part of the imperial family.

Here, first, are three documents that, though only indirect testimonies, do raise the question of the transfer of the Romanovs from Ekaterinburg and the identity of those who accompanied them. All three are contemporary with the events to which they refer. The first, discovered by Summers and Mangold, emanates from a British agent who was present in Ekaterinburg. The two others, which appear in the book edited by Ross, are from the file compiled by Sergeyev before he was removed from the case.

All three documents show that a transfer actually took place, but without indicating of whom. All three leave open the possibility that the Tsar alone was executed – with his staff, that is, but without his wife or his daughters. From the first document we also learn that the Romanovs' hair had been cut, as though to disguise their appearance. Were they all being prepared for a departure? We recall Nicholas's exclamation, in the Bolsheviks' account: 'So we are not being taken anywhere?'

Here, first, is the evidence that Summers and Mangold found in the British public records. It comes from Sir Charles Eliot, High Commissioner for Siberia:

The position of the bullets indicated that the victims had been shot when kneeling and that other shots had been fired into them when they had fallen on the floor ... There is no real evidence as to who or how many the victims were but it is supposed that they were *five*, namely, the Tsar, Dr Botkin, the Empress's maid and two lackeys. No corpses were discovered ... but it was stated that a finger bearing a ring believed to have belonged to Dr Botkin was found in a well.

On 17 July a train with the blinds down left Ekaterinburg for an unknown destination, and it is believed that the surviving members of the imperial family were in it ... It must ... be admitted that since the Empress and her children, who are believed to be still alive, had totally disappeared, there is nothing unreasonable in supposing the Tsar to be in the same case.

It is the general opinion in Ekaterinburg that the Empress, her son and four daughters were not murdered but were dispatched on 17 July to the north or west. The story that they were burnt seems to be an exaggeration of the fact that in a wood outside the town was found a heap of ashes, apparently the result of burning a considerable quantity of clothing. At the bottom of the ashes was a diamond and as one of the Grand Duchesses is said to have sewn a diamond into the lining of her cloak, it is supposed that the clothes of the imperial family were burnt here. Also hair, identified as belonging to one of the Grand Duchesses, was found in the house. It therefore seems probable that the imperial family were disguised before their removal. At Ekaterinburg I did not hear even a rumour as to their fate, but subsequent stories about the murder of various Grand Dukes and Grand Duchesses cannot but inspire apprehension.

The second document is contradictory and imprecise in that it implies that only the Tsar died and that a witness considered that the poster announcing this event was not to be taken seriously:

Testimony, 13 September 1918, of Fyodor Ivanovich Ivanov
I have a hairdressing salon in the new station at Ekaterinburg and I remember very well that a day or two before the Bolsheviks announced the execution of Nicholas the station commissar Gulyaev told me that he had a lot of work to do.

'What sort of work?' I asked.

'Today we are taking Nicholas away ...'

As there were other people present I did not dare ask him where to. That evening, when Gulyaev came again to my salon, I asked him about what he had said, as there was no train waiting in the station, and he replied that the train would leave from Ekaterinburg's other station, but he gave me no details.

Next day, seeing the Red Army commissar Kucherov, I asked him:

'Is it true that the Tsar has left from the other station?'

'Yes, that is so.'

'And where has he gone?'

'What business is that of yours?'

When I met Gulyaev that same day at the station I asked him what had happened to Nicholas.

'He's already *khalymuz* [*kaput*?].'

'What does that mean?'

He replied: 'It's finished,' and by this I understood that Nicholas had been killed ...

Two days later, when I met them both in the station buffet, I asked them what the poster meant, and they answered:

'What's it matter what they write?'

I asked the sailor Grigory, who often slept in the men's room at the station:

'So, they've shot them?'

'I doubt it.'

'Have they taken them away, then?'

'They left the town alive.'

But he did not tell me where they had gone.

There was great secrecy among them as to what had happened to Nicholas and during that time they were all very agitated. Nobody said anything about the family and I was afraid to ask too many questions.

F. I. Ivanov did not know where the train was bound for.

The third testimony, which is also missing from Sokolov's report, mentions the train's destination and also speaks of the Tsarevich's death:

I had an intimate involvement with K. Kanevtsev, of the Cheka ... I was repelled by him but surrendered myself physically. Because I was not interested in their Bolshevik affairs, I never asked him about their secrets. I remember that about two days before the announcement of the murder of the former Tsar, Kanevtsev came round to my apartment at about 4 p.m. and told me that the Bolsheviks had killed the former Emperor; it seemed to me that as he said this Kanevtsev had tears in his eyes, and he sort of turned away. He told me, when I asked him, that they had buried him – out in the open space – and that he had 52 bullets in him. He told me that the family had left for Nevyansk. He said that the Tsarevich was dead. Next day he went away, to Perm, he said, to fetch gold ...

Signed by Zinaida Andreyevna Mikulova, about 9 August 1918.

The following testimonies mention the presence of the family at Perm and refer in passing to an escape by Anastasia:

Testimony by Natalya Vasilevna Mutnykh (8 March 1919)

I found out by chance that the family of the former sovereign Nicholas II – his wife and four daughters – were transferred from the town of Ekaterinburg to Perm, where they were kept very secretly in a cellar in the Beryozin house, where there was a workshop. One of the girls escaped from this cellar in September, but was caught somewhere beyond the Kama and brought back, while the family were transported elsewhere ...

I was interested in the presence of the Tsar's family in Perm and, making use of the fact that my brother Vladimir Mutnykh had to go on duty at the place where the imperial family were being kept, I asked him to take me with him and show me them. My brother agreed and we set off. This was in September. At Beryozin's rooms we went down to the basement and I saw the room where, in the poor candlelight, I could make out the former Empress Alexandra Fyodorovna and her four daughters, who were in a terrible state but I recognized them only too well.

With me was Anya Kostina, Zinoviev's secretary. She left later for Petrograd. The Emperor's family was hidden in a barracks somewhere in the country.

(cf. Ross, document no. 116)

Anastasia's attempt to escape led to a big chase. She was caught, beaten, probably raped and brought back to the basement. Here is the testimony of the doctor who was summoned urgently by the Cheka:

Testimony of Dr Utkin, 10 February 1919

At the end of September 1918 I was living in Perm on the corner of Petropavlovskaya and Obvinskaya Streets, in the building of the Peasant Land Bank ... At the end of the first half of September the Cheka began to occupy the building ... Shortly after 20 September, at about five or six o'clock in the evening, an orderly came to me from the Cheka and said: 'Doctor, go at once to Malkov' [Malkov was the head of the local Cheka. Utkin was taken to a room where] ... a young woman was lying, semiconscious, on a divan. She was plump and her dark brown hair had been cropped close. Beside her were some persons among whom were Vorontsov, Malkov, Trofimov, Lobov and others whom I did not know. With these men there was also a woman, about 22–24 years old, blonde, of average height and build. At my request all the men left the room. The woman remained, saying that, as she was a woman, she would not be an embarrassment. As a doctor, I was well aware that she was acting as an informer.

When I asked the sick woman 'Who are you?' she raised her head and said, in a tremulous, distressed voice: 'I am the sovereign's daughter Anastasia.' Then she fainted.

The sick woman looked like this. All round her right eye her face was swollen and there was a laceration about one and a half centimetres long at the corner of her mouth. I did not notice any injuries to her head or chest. I was forbidden to examine her sexual parts when I began to do this. I then put a bandage on her and prescribed some medicine. I was then asked to leave.

At about ten o'clock in the evening I went again, on my own initiative, to see the patient. She wandered in her speech, uttering disconnected words and phrases. After that visit I did not see her again. When I had put the bandage on her she said: 'Doctor, I am very, very grateful to you.'

The mention of 'hair cropped close' corroborates what we know of the haircut given to the Tsar's daughters. The doctor says that he was not allowed to examine the patient's sexual parts. No doubt she had been raped: I shall come back to this point. There is a second testimony by Dr Utkin in the Houghton Library.

Testimony of Dr Utkin, 14–15 June 1919

After the interrogation (in February) I went to the pharmacy where I had prepared the medicines. These prescriptions were in the pharmacy of the provincial zemstvo, where I had received them from the person in charge, Korepanov ... I remember that when I was writing out the prescription I wondered: should I write the name 'Romanova', or not? When I asked the Bolsheviks about this they told me to put just a letter, any letter of the alphabet. And I put the letter N. at the top. That was why the prescription was left there and not entered in the book.

Later, Dr Utkin was shown a series of photographs, and he identified in them both Anastasia and the young woman who acted as a spy. Kirsta said to him: 'Yes, doctor, we know that she is here. We now have all the threads of the story. Anastasia Nikolaevna was brought here from the district on the other bank of the Kama.'

When he was about to sign his deposition, Dr Utkin recalled that he had originally made a mistake, which he corrected; she had not said: 'I am the Emperor's daughter Anastasia' but 'I am the sovereign's daughter Anastasia.' Utkin also commented that he thought she was mentally deranged. That may very likely have been the case if she had been beaten up, and even perhaps raped. She was not yet eighteen.

Utkin's testimony was not retained by Sokolov and Dieterichs, as they considered the doctor to be a nervy person and not dependable. No doubt he was nervy, since they began by asking him why he had not come forward with this story immediately after the liberation of Perm by the Whites.

According to the evidence given by the woman Mutnykh, one month later the young Anastasia (or Tatiana, she was not sure) was taken to Glazov, and then in the direction of Kazan. Mutnykh's brother Vladimir said that she knew the bodies of the ex-Tsar and his heir had been burned at Ekaterinburg. 'She [Anastasia] died later, whether from injuries or maltreatment I don't know. But I know that she was buried at one in the morning, not far from the racecourse, and the Bolsheviks treated this as a big secret.'

This testimony provides some pointers to the fate of Alexei and also to what could be the 'second death' of Anastasia. The burial at one in the morning suggests, though, that the reference is to the nocturnal doings near Ekaterinburg. The brittleness of the testimony in this respect inspires caution.

We note, particularly, that the trend of the investigation and of the depositions leans this way or that according to who is doing the investigating: the military, Sergeyev, Sokolov or the police. The unavowable hypothesis does, however, seem to have more substance to it in the light of these materials from Perm; for the documents concerned, which could support it, do not come merely from one of the files of Sokolov's investigation but also from other sources that may enable us to check that investigation's reliability.

The materials of the investigation include a deposition by Evgeniya Sokolova, a teacher at Perm, who declared, on 17 March 1919, that 'The Tsaritsa and her three daughters left Perm by train, after September.' What is interesting here is that she speaks of three daughters only, not four, which fits in with other information.

The vital document that corroborates it comes from Alexis de Durazzo, Prince of Anjou, the grandson of a lady who claimed to be Maria, one of the ex-Tsar's daughters. She left a handwritten testimony dated 10 February 1970, 'to be opened after ten years', which Alexis published in 1982. 'On 6 October 1918, in the morning, in the town of Perm, where we had been since 19 July, my mother and my *three* [sic] sisters, who had been separated, were taken to the train. I arrived in Moscow on 18 October. G. Chicherin, Count Czapski's cousin, handed me over to a Ukrainian representative, to leave for Kiev.'

This manuscript, written in 1970 by the trembling hand of a woman of seventy-one, is thus the work, according to her grandson, of a lady named Maria who claimed to be one of the ex-Tsar's daughters. He adds some oral testimony by his grandmother, who, 'out of considerations of safety', had for a long time kept silence. It gives information about the stay at Perm, the separation of the Tsaritsa and her four daughters into two groups, with Maria being kept in the Beryozin house along with Anastasia, 'who disappeared on 17 September: she escaped'. For the second time, then; but there was no further news of her.

The testimony of 'Maria' ties together the different threads of the affair. Having returned from Ekaterinburg to Perm, Beloborodov told her a few days later that they were 'going to Moscow' and she should get herself ready. They were to travel in groups: 'No luggage, just a small case or a bundle.' On 6 October 1918 they were taken, on foot, to Perm railway station. The Bolsheviks agreed to the Tsaritsa's request to keep Tatiana with her. Olga looked at Maria and said, in English: 'What does it matter now? Nothing worse can happen. God's will be done.' She got into one train and Maria into another. Maria told how, during the journey, some brute ordered her to remove her ear-rings. She would not do this and he took them by force, which left her with a scar. Maria reached Moscow on 18 October and was lodged in the former residence of Bruce Lockhart, where Lunacharsky's wife received her. Chicherin soon appeared, courteous, kissing her hand and saying that 'the foreign embassies were arranging for her departure and that of her family'. She was to leave for Kiev.

'We Communists', said Chicherin, 'overthrew the tyranny of your family, but we respect human life.' He added that she was being entrusted to the Ukrainian government, 'who are, of course, puppets', but that at Kiev there were 'representatives of your German family, and you must go there'. Shortly after this, Hetman Skoropadsky sent a special train, and Maria went off in it, holding a passport in the name of Countess Czapska, the Polish Count Czapski being a cousin of Chicherin's (cf. Durazzo, *Moi, Alexis ...*, pp. 180–3).

In that same week, on 23 October 1918, Karl Liebknecht was released from prison by Wilhelm II and Chancellor Max von Baden. Later, Jogiches, another Spartakist, of Polish origin, was released. Was this a coincidence? Or was it the first exchange of hostages between East and West? We do know for certain that, during that summer, Karl Radek, who along with Ioffe represented the Bolshevik authorities in Berlin, had taken the initiative in proposing such an exchange: the Tsaritsa and her daughters to be freed in return for the freeing of the ultra-leftists imprisoned by the Kaiser. He had assured the Germans that the Romanov women were safe.

This testimony by Alexis's grandmother and that by Sokolova are the only pieces of evidence we possess that speak explicitly of a departure from Perm and then a move to Kiev. They do, however, fit in with other documents, from the Spanish, German and Vatican archives, which refer to negotiations concerning such a transfer, though these do not prove that the transfer actually took place. They all date from August and September 1918, when it had already been announced officially that the Tsar had been executed, and that the same fate had befallen all the Romanovs who were interned at Alapayevsk, north of Ekaterinburg.

We can imagine the concern felt in the courts of Europe, those of Berlin, Copenhagen and London. The court of Madrid (Spain being neutral in the war that was still going on) sought to intervene to persuade the Bolsheviks to free the Tsaritsa and the princesses, if they were still alive. The King of Spain was especially concerned. Alfonso XIII was related, through his wife, to Queen Victoria and so to Alexandra.

The historian Carlos Seco Serrano has found, in the Spanish public records, that on 4 August 1918 the Spanish court thought that the Tsar had been murdered but his wife and daughters were still alive. Here is the relevant document.

4 August 1918. Letter from the Spanish Ambassador in London, Alfonso Merry del Val, to the Minister of Foreign Affairs, Eduardo Dato.

The interruption of our conversation yesterday prevented me from putting to Your Excellency an idea of definite importance and urgency, in connection with the approach you have undertaken on behalf of the widow and daughters of the unfortunate ex-Emperor of Russia.

He goes on to mention Nicholas's mother, the Dowager Empress:

Would it not be possible to include the case of this august lady in the proposed negotiations? She is, as you know, a sister of Queen Alexandra, the mother of King George V, and an intervention in her favour would make more acceptable to the British royal family and to British public opinion the move being prepared to obtain the release of Empress Alice. The Empress . . . is much disliked . . . she is regarded as an agent, conscious or unconscious, of Germany and as bearing the principal responsibility, even if unintentional, for the revolution, through the bad advice she gave to her husband, whom she dominated absolutely . . . The very strong feeling . . . against Empress Alice is such as to exclude all possibility for her to reside in the United Kingdom.

The second document, dated a month later, relates to the negotiations with Chicherin. '6 September 1918. Telegram 858.' (Entrusted by Madrid with the task of negotiating the transfer of the Empress and her children, Fernando Gomez Contreras left Petrograd for Moscow accompanied by the chargé d'affaires of the Netherlands. He had two interviews with Chicherin, on 1 and 5 September.)

The People's Commissar received us, an hour later than the appointed time, in the dirty building that serves as the Ministry of Foreign Affairs. He was accompanied by another Jew, his assistant (*accompañado de otro israelito es su adjunto*).[6] I explained to him the humanitarian wish of our sovereign, saying that there is no question of intervening in Russia's internal affairs and that the imperial family would stay in Spain and keep out of politics. The commissar began by showing dissatisfaction because we had come to intercede on behalf of persons who had caused so much harm to the people. He asked me, in sharp terms, to recognize the Soviet power officially, adding that, for this question to be settled, it was necessary for the two sides to accord each other recognition. He went on to say that, for this reason, he doubted the validity of our assurances to the effect that

6. Karakhan is meant. He was not, in fact, a Jew (any more than Chicherin was), but an Armenian. He was executed in the Stalin period.

the imperial family would remain aloof from any counter-revolutionary movement. Alluding to the detention [in 1916] of Trotsky in Spain (*alude a detención de Trots en España*), he claimed that our country would be transformed into a centre of reaction and counter-revolution against the international proletariat.

We emphasized ... that the pointless martyrdom of this defenceless woman would bring upon them the reprobation of the entire world ... After a painful discussion and much effort, I secured his promise to submit our request to the next meeting of the Central Executive Committee.

On 15 September, referring to the instruction he had received on 22 August, 'to request the transfer of the imperial family to Spain', Gomez Contreras adds, speaking of Chicherin (and Karakhan), that he has said that 'he will see to it that a solution is found to the situation of the imperial ladies in the sense of their being set free'.

Some data from Russian or German sources relate to these negotiations. In his diary, Milyukov, the former Minister of Foreign Affairs, who was in Kiev in the summer of 1918, mentions that he had talks with the Germans in which he spoke of marrying the Tsar's daughter Olga to Grand Duke Dmitri Pavlovich in connection with a plan to create a Ukrainian state 'protected' by Germany and Russia.[7]

In the public records in Berlin, Summers and Mangold found evidence of statements made by Count Alvensleben, who was at the centre of the negotiations between the Germans, Russians and Ukrainians. He was 'the eye of Wilhelm II', said Jean Pélissier, a French journalist in Kiev. On 5 July 1918 Alvensleben told General Dolgorukov that 'between 16 and 20 July a rumour will be put about that Nicholas II is dead, and this will be untrue'. And when the

---

7. In a letter of 26 June/1 July 1918, sent from Vologda, Grand Duke Nicholas Mikhailovich (who was executed on 28 January 1919) wrote to Frédéric Masson that the Dowager Empress had been liberated by the Germans from the Bolsheviks on 1/14 May, but she had declined to go to Denmark under their protection, saying that she would prefer to be killed by Russians.

communiqué duly appeared, a requiem for the Tsar was held in Kiev, and another in Copenhagen, where, having heard of what was being said in Kiev, the (anti-Bolshevik) Russian Ambassador to Denmark told the French ambassador that this was a false story being circulated in order to save the Tsar's life.

Did the Germans hope, as the Kaiser had demanded, to save the whole family, the ex-Tsar included? We may suppose that this was so; but in Ekaterinburg the Bolsheviks had decided to execute Nicholas and evacuate only the women, resorting to all this camouflage so that the Left SRs should imagine that the whole family were dead. The Tsaritsa's brother Ernst of Hesse had been to see Wilhelm II and, in the opinion of Nicholas Mikhailovich in his letter of 26 June/1 July, 'orders were sent from Berlin to Lenin and Trotsky concerning the sovereign and his family'. Later on, Ernst, who naturally sought to obtain news of his sister, sent through an intermediary a message to the British court in which it was said that 'Ernie now telegraphs that he has heard from two trustworthy sources that Alix and all the children are alive.' This telegram, discovered by Summers and Mangold, is dated 27 September 1918 – more than two months after the presumed death of the whole family.

There is another relevant source, namely the archives of the Vatican. A letter from the Minister of Foreign Affairs in Berlin, dated 21 September 1918, to His Eminence Cardinal von Hartmann, Archbishop of Cologne, says that 'the Russians have assured [the Germans] that they will not interfere in their affairs, that they will protect the Grand Duchesses from the people's fury and that they are planning to move them to the Crimea'.

The Vatican was in fact also playing its part in this extensive enterprise aimed at saving the wife and daughters of the ex-Tsar. On 10 October 1918 – that is, after the move from Perm to Moscow, if that did indeed take place – the Bolsheviks replied for the first time to the Holy See, through the Austro-Hungarian consul in Moscow 'that they did not know where the Tsaritsa and her daughters were'. They repeated this claim on 28 October. A diplomatic lie: of course, if the women in question were now 'somewhere in the

Ukraine', the Bolsheviks could indeed say that they did not know just where they were.

Another very curious document refers to the entire family leaving the Ukraine in 1919. It is a message to the King of England. The writer speaks of the Tsar as though he were still alive (unless there has been an error in the transcription). The interest of the document lies in the fact that, although Olga, Tatiana and Maria are mentioned explicitly as participating in the journey, nothing is said of Anastasia.

From Lord Hardinge of Penshurst, Permanent Under-Secretary for Foreign Affairs [handwritten date: 3 or 5 June 1919]

The King

Your Majesty,

In response to Your Majesty's enquiry, I have ascertained from the Chargé d'Affaires in Vienna that the route taken by His Imperial Majesty the Czar and the Grand Duchesses Olga, Tatiana and Maria was as you were informed by Her Majesty the Queen Mother, from Odessa–Constantinople, arriving February 26th

From Constantinople by train arriving Sofia February 28th

From Sofia to Wien on March 3rd arriving Wien March 7th

From Wien to Linz by car arriving 8th

From Linz to Wroclaw or Breslau on May 6th arriving Wroclaw May 10th.

I am your Majesty's obedient servant.

Hardinge

It is strange that the Tsar should be mentioned here. However, we have seen that, on some earlier occasions, it was believed that he too had left Ekaterinburg for Perm. Was the Foreign Office still unaware of what was going on between Moscow and Berlin? Was this letter meant to mislead? In any case, it brings up the problem of Anastasia.

Regarding Anastasia, if the journey from Perm to Kiev did actually happen, it seems that one possible hypothesis might be as follows.

Transferred to Perm with her sisters, she escaped in company with one of her young captors. Caught, beaten and/or raped by the soldiers, and brought back, she was examined by Dr Utkin. Then she disappeared again, we do not known how. In any case, she was not at Perm railway station for the journey to Kiev, and after September her sisters did not know what had happened to her. Herself unaware that her sisters had left on 6 October, Anastasia might have said and believed, later on, that she was the only one who 'escaped'. She nevertheless also claimed that 'nothing happened at Ekaterinburg in the way that has been said'. She was unable to say anything more about herself – in my opinion because, eighteen years old, traumatized by rape and injuries, she could never express it, 'confess' it, owing to a sort of guilt feeling intensified by her Victorian education and her sense of rank. Perhaps she may also have felt some shame for having 'abandoned' the others in order to escape alone.

Was her companion German or Austrian? She turns up, pregnant, in Germany. When she reappears in 1919 she is dying, and several members of her family recognize her, in particular her two aunts, Olga and Xenia, the sisters of Nicholas II. But who, at that date, knew what had become of her sisters? Ernst of Hesse, perhaps, Alexandra's brother, who had caused that telegram to be sent, on 27 September 1918, stating that he knew they were alive. But we have no other information.

Now, while her mother and sisters may have been looked after, in Kiev or Podolia, Anastasia was a nuisance. Contrary to expectations, she recovered, became Mrs Tschaikowski and, believing herself to be the sole survivor of the imperial family, laid claim to the Romanov inheritance, which Grand Duke Cyril Vladimirovich and her aunts Olga and Xenia intended to take over. On this point what Gleb Botkin says is convincing. Being the son of Dr Botkin, the Tsar's physician who was murdered at Ekaterinburg, he knew the sisters well, and was their playmate for several years, right down to their incarceration at Ekaterinburg. He recognized Mrs Tschaikowski at once as Anastasia. And, according to him, her aunts Olga and Xenia altered their attitude, saying that they would not recognize her as Anastasia until a court had so decided, only when she had recovered and asserted her claim to the inheritance.[8] Backed by Cyril and Ernst of Hesse, they now started to talk about an act of imposture ... And Anastasia, under questioning, said she remembered that Ernst had come to visit his sister the Tsaritsa during the war, whereas he was then a general in the German Army: a blunder.

The other sisters had been more discreet. Their timidity ensured their survival. Besides, the traumatism they had suffered, the insecurity that prevailed throughout central Europe in 1919, their fear of being discovered by the Bolsheviks and murdered like others of their family – all these conditions suffice to explain why they stayed hidden. Maria, apparently, did not decide to reveal the truth about herself until decades later. According to Alexis de Durazzo, Anastasia was sacrificed to dynastic calculations; Cyril having proclaimed himself head of the House of Romanov, Anastasia was an embarrassment. People began to mutter about the likelihood that all four girls had been raped, which would have been a real dishonour to the family. The other sisters were helped by the Queen of Romania and by the Kaiser during his exile in Holland.

All these facts fit together, more or less. True, we have seen that Maria mentions her three sisters in her written testimony and that in her oral testimony she mentions the disappearance of Anastasia. But we note also that the Bolshevik leaders connected with these events were precisely those who spoke openly about the Romanov girls and their mother having survived: Chicherin, first and foremost, whose role was central, in consequence of the position he held in 1918; then Zinoviev, whose secretary had been, we are told, in Perm at the time of the transfer; and Litvinov, who was one of Chicherin's assistants and who made his statement about the women's survival in December 1918, after the Armistice, when there was no longer any need to deceive the Germans on this issue.

The point is important because some – the Russian anti-Bolshevik

8. Grand Duke Andrew had recognized her, too. When Felix Yusupov heard that Gleb Botkin had done the same, he went to see him with a view to persuading him to drop Anastasia 'and go over into the camp of his mother-in-law, Grand Duchess Xenia'.

historian Melgunov in particular – have supposed that the negotiations between the Bolsheviks and the Germans were a swindle all along, in that the Bolsheviks talked about releasing the women when they knew very well that they were dead.

We do possess consistent information regarding both Anastasia and Maria, and this information coincides fairly well, though that does not constitute irrefutable evidence, since Maria's testimony was published only later. Alexis de Durazzo produced his evidence after reading Summers and Mangold's book, which cleared a path for him. But he had not read *Gibel' Tsarskoy Sem'i*, which appeared in 1987.[9] And the thinness of our information regarding the Tsaritsa and Tatiana renders these claims of survival only hypothetical. Alexis writes that the Tsaritsa had been given refuge in a convent in Podolia, but says nothing about Tatiana except that she corresponded with his grandmother. It is true that some unpublished information about Maria and Anastasia is available. A direct testimony by Prince Ghika shows that, in 1920, he knew about the marriage of Grand Duchess Maria to Prince Dolgoruky. He also testifies that Grand Duke Cyril asked Queen Marie of Romania – granddaughter of Alexander II and of Queen Victoria – 'to say nothing more about the arrival in Bucharest of the two Grand Duchesses' in 1919, 'for family reasons'. In her book about Marie of Romania, *The Last Romantic*, Hannah Pakula mentions that, also in 1919, the British court told the Queen that Maria Pavlovna, the daughter of Grand Duke Paul, could not be properly received if she came to London. This is explicable by a desire on the part of the British royal family not to be reminded of the behaviour of George V in 1917–18, even though the King had later helped to finance a plan for getting the imperial family out of Russia by the northern route.

The Romanian Queen said that even 'if all the world meant to behave like selfish beasts', *she* would never deny her exiled Russian relatives. And a Soviet diplomat said something similar to Botkin Junior when he was in the United States: the Romanovs acted more cruelly towards Anastasia than the Bolsheviks did.

There is also a photograph of Maria and Olga together, taken,

9. I questioned him at length in Madrid in March 1989.

however, in 1957. Does it really show them? And another testimony: Sister Pascalina Lehnert, Pius XII's servant, states that this Pope received the two princesses, Olga and Maria ('it really was them') at the Vatican, but at a date she cannot be sure of – some time, evidently, between 1939 and 1957.

All the above leads us to ponder on the question of how Nicholas II really died, and why mystery surrounds this question.

The 'unavowable' hypothesis presumes a different death for Nicholas II. Let us, therefore, go back to mid-July 1918. The version given by Alexis de Durazzo, on the basis of what his grandmother told him, describes the events like this:

On 23 June 1918 [6 July, New Style], the Emperor was taken from the Ipatiev house by Yurovsky in order to have certain conversations ... He met two persons who had come from Moscow ... He was offered the chance to go abroad if he would agree to certain conditions. Out of concern for his family, the Emperor agreed ... He was brought back to the Ipatiev house on 9 July ... On 12 July Yurovsky told the imperial family to get ready for a long journey that was to take place in the greatest secrecy. For this purpose he asked them to alter their appearance ... On 15 July he told the Emperor that he was to go first, with Alexei, separately from the rest of the family. My grandmother [Maria] was quite clear on that point ... They [Nicholas and Alexei] were taken away during the night of 15 July ... My grandmother did not know where they were being taken to.

Corroboration for this story is provided by the fragments of cut hair, from heads and a beard, which were found in the Ipatiev house. The young woman whom Dr Utkin examined had short hair, too, though all four girls had had their hair cropped once already, probably a year earlier, as we see from a photograph.

What happened then? Only the testimony of T. Chemodurov, which is included in *Gibel' Tsarskoy Sem'i* (he was a servant who escaped from the massacre), provides a detailed account of subsequent events. This was published by Carl Ackerman, correspondent of the *New York Times*, in *Vestnik Man'chzhurii*, No. 31.

Testimony of Parfen Alexeyevich Domnin,
real name Terenti Ivanovich Chemodurov

Beginning with the first days of July, aeroplanes began to appear nearly every day over Ekaterinburg, flying very low and dropping bombs, but little damage was done. Rumours spread about the city that the Czechoslovaks ... would shortly occupy the city.

One day the former Tsar returned to the house from his walk in the garden. He was unusually excited, and after fervent prayers before an icon of St Nicholas the Wonder-Worker he lay down on a little bed without undressing. This he never did before.

'Please allow me to undress you and make the bed,' I said to the Tsar.

'Don't trouble, old man,' the Tsar said. 'I feel in my heart I shall live only a short time. Perhaps today, already—' But the Tsar did not end the sentence.

'God bless you, what are you saying?' I asked, and the Tsar began to explain that during his evening walk he had received news that a special council of the Ural district soviet of workers', cossacks' and Red Army men's deputies was being held to decide the Tsar's fate.

It was said that the Tsar was suspected of planning to escape to the Czech army, which was advancing towards Ekaterinburg and had promised to wrest him from the Soviet power. He ended his story by saying resignedly: 'I don't know anything.'

The Tsar's daily life was very strict. He was not permitted to buy newspapers, and was not allowed to walk beyond the limited time ...

The former heir to the imperial throne, Alexei Nikolaevich, was ill all the time ... One evening Alexei came running into the Tsar's room, breathless and crying loudly, and, falling into his father's arms, said, with tears in his eyes: 'Dear Papa, they want to shoot you.'

The Tsar whispered: 'It's the will of God in everything. Be quiet, my son ... Where is Mama?'

'Mama is crying,' said the boy.

'Ask Mama to calm herself ...'

The Tsar knelt before the icon of St Nicholas, praying for a long time ...

On 15 July, late in the evening there appeared suddenly in the Tsar's room the commissar of the guard, who announced: 'Citizen Nicholas

Alexandrovich Romanov, you will follow me to the Ural district soviet of workers', cossacks' and Red Army men's deputies.'

The Tsar asked, in a pleading tone: 'Tell me frankly, are you leading me to be shot?'

'You must not be afraid ... You are wanted at a meeting,' the commissar replied, with a smile.

Nicholas Alexandrovich got up from his bed, put on his grey soldier's blouse and his boots, fastened his belt and went off with the commissar.

... Nicholas did not return for a very long while, about two hours and a half at least. He was quite pale, his chin trembling.

'Old man, give me some water,' he said.

I brought him water at once. He emptied a large cup.

'What happened?' I asked.

'They have informed me that I shall be shot within three hours.'

Chemodurov goes on to explain:

During the meeting of the Ural district soviet a minute ... was read in the presence of the Tsar ... It stated that a counter-revolutionary plot had been discovered, the plotters being a secret organization calling themselves the Association for the Defence of Our Native Country and Freedom. The object of the plot was to suppress the workers' and peasants' revolution by inciting the masses against the soviet by blaming it for all the hard consequences resulting from imperialism all over the world – war and slaughter, famine, unemployment, the collapse of transport, the advance of the Germans, and so on.

... The counter-revolutionists were attempting to unite all the non-Soviet political parties, socialists as well as monarchists ... The evidence presented showed that at the head of the plot stood a personal friend of the Tsar, General Dogert ... In this organization were working also Prince Kropotkin, General Staff Colonel Eckhart, Engineer Linsky and others. There were reasons for believing that Savinkov was also in direct connection with this organization and that he was intended to be the head of the new government, as a military dictator ...

The testimony stated that during the last few days a new plot had been discovered, having for its object the rescue of the former Tsar with the help of General Dutov.

... In view of this evidence, together with the difficult situation caused by the decision of the Ural district soviet to evacuate Ekaterinburg, the former Tsar was to be executed without delay ...

'Citizen Nicholas Romanov,' said the chairman of the soviet to the former Tsar, 'I inform you that you have three hours to write your last letters. Guard, do not let Nicholas Romanov out of your sight.'

Soon after Nicholas returned from the meeting his wife and son came to see him, weeping. Often Alexandra fainted and a doctor had to be called. When she recovered she knelt before the soldiers and begged for mercy. The soldiers answered that mercy was not within their power.

'Be quiet, for Christ's sake, Alix,' the Tsar repeated several times in a low tone, making the sign of the cross over his wife and son.

After this, Nicholas called me and kissed me, saying: 'Old man, do not leave Alexandra and Alexei. You see, there is nobody with me now. There is nobody to look after them, and I shall soon be taken away.'

Later it turned out that nobody except his wife and son, of all his loved ones, was permitted to bid farewell to the former Tsar. Nicholas and his wife and son remained together until five more soldiers of the Red Army appeared, along with the chairman of the soviet and two of its members, both workers.

'Put on your overcoat,' the chairman ordered ...

Nicholas, who did not lose his self-possession, put on his coat, kissed his wife and son and me again, made the sign of the cross over us and then, addressing the men, said in a loud voice: 'Now I am at your disposal.'

Alexandra and Alexei fell to the floor in a fit of hysterics. I made an attempt to bring mother and son to, but the chairman said: 'Later. There must be no delay. You can do that after we have gone.'

'Allow me to accompany Nicholas Alexandrovich,' I asked.

'No accompanying,' was the stern answer.

So Nicholas was taken away, nobody knows where, and was shot during the night of 16 July, by about twenty Red Army soldiers.

Before dawn the next day the chairman of the soviet again came to the room, accompanied by Red Army soldiers, a doctor and the commissar of the guard. The doctor attended to Alexandra and Alexei. Then the chairman said to the doctor: 'Is it possible to take them away immediately?'

When he answered 'Yes', the chairman said: 'Citizeness Alexandra Fyodorovna Romanov and Alexei Romanov, get ready. You will be sent away from here. You are allowed to take only what you can't do without, not over 30 or 40 pounds.'

... Mother and son soon got ready.

... While they were making their preparations, the chairman said to me: 'And now, old man, you clear off, nobody is staying behind here, there's nothing more to be done.'

The Empress and her son were taken away in an automobile, I don't know where to.

What is strange here is that the author of the testimony, Parfen Alexeyevich Domnin, never existed. This was a pseudonym for Chemodurov, presumably given him by the American journalist for his own safety. That, at any rate, is the hypothesis put forward by Summers and Mangold. The account given 'fits' fairly well with the other pieces in the jigsaw puzzle. Summers and Mangold adduce other testimony, too, that supports what is said here. The Tsar's old servant, if it really was him, was said to have been ill and under care in the prison hospital for several weeks. He could not therefore have been on the spot, unless he went out for a walk and then returned to the hospital, thus escaping the fate suffered by the Tsar's other servants. He testified later, before Judge Sergeyev, that the hairs found in the Ipatiev house were from the heads of the four princesses, whose hair had been cut short, and from the beard of the ex-Tsar, which Nicholas had half cut off, as though the whole family had begun to try to alter their appearance. Yet neither Sergeyev nor Sokolov included this testimony in their reports. Was this because the witness was considered not to have been present at the moment of the crime? The testimony is, in any case, confirmed by a passage in *Pravda*, stating that the Tsar had been shot *outside the town*, by ten Red Guards.

What is surprising is that Ross, who is usually very precise in the notes he supplies, says merely that Domnin never existed, without even mentioning the proof given by Summers and Mangold that 'Domnin' was really Chemodurov, though he knows and quotes from their work. He demolishes the reliability of Domnin's testimony solely by pointing out that there were no Czechoslovak or Russian aeroplanes operating in Siberia at the relevant time. He does not include Captain Malinovsky's speculation, and says that Sergeyev believed that the whole family were killed together. Furthermore, since none of his documents is later than 21 February 1920, he does not include what Alvensleben said to Dolgorukov on 5 February 1921. Finally, Ross attaches no importance to the evidence concerning Perm, but merely remarks that Dr Utkin is the only person who claims to have spoken with the self-styled Anastasia. In short, although he publishes documents that Sokolov left out, Ross considers these documents to be of little significance and not sufficiently supported to justify challenging the accepted version of what happened.

Yet the documents that point to the whole family having been killed are also contradictory. None of these accounts coincide. In some the bodies are burned, in others buried, or else the heads are cut off – and now we are shown alleged photographs of all the skulls. Sometimes the assassins are Chekists, sometimes soldiers; sometimes they number four, sometimes eleven. The victims do not enter the room in the same order in every account. And so on. Furthermore, it is surprising that, for a crime committed in secret, nearly 200 witnesses were summoned in the White-occupied zone, when the Reds – among them, presumably, the killers – had withdrawn westwards.

Many of the depositions include precise and concrete information, as though the witness had actually been present. These testimonies contradict one another, though – in one case arguing for the murder of the whole family, and in another showing that some of the family survived, or even that everybody, the Tsar included, survived.

We have seen that the first investigators thought that there had been an attempt at deception; that Judge Sergeyev also thought that only the Tsar and his son had been executed; that the military police, under the Czechoslovaks' command, brought proof that the Romanov girls were still alive at Perm, but that General Dieterichs removed them from the investigation. While Dieterichs appointed Sokolov in place of Sergeyev, who was considered to be an SR and unreliable, in view of his conclusions, Kirsta was taken off the case because his superior, General Gajda, clashed with him. We have also seen that all whose testimony contradicted, more or less, the thesis that the whole family had been killed themselves died tragic deaths.

Yet the evidence pointing to the contrary is not so very convincing either. It leaves unanswered the question of the death of Alexei. Nevertheless, it does possess the advantage of coherence. But against it is ranged the power of the apparatuses, both White and Red. What is at stake for them here?

The zeal of the Whites to affirm that the entire imperial family were murdered is due not only to dynastic interest but also to another, independent reason.

Of course, the Romanovs who gathered round Cyril when he proclaimed himself head of the House of Romanov could not endure the idea of any of the sisters – whether Anastasia or another – having survived, since that would have spoiled the prospects of their branch. But the Whites, especially Kolchak, had another reason for taking that line. After 1917 Kolchak was no longer a monarchist, any more than Kornilov was. He wanted to restore the old order, but minus a Tsar. While it is true that some of the military leaders and of the Cadets and Octobrists supported, after the end of 1916, the idea of a palace revolution in favour of a different Romanov, the situation was completely changed for them by the February Revolution. Apart from Milyukov and a few others, these counter-revolutionaries abandoned the idea of a constitutional monarchy, no matter with what branch of the Romanov family at its head. They were anti-Bolshevik but, now, opposed to a restoration, counting rather on the establishment of a military dictatorship or perhaps a Fascist-type regime.

In view of the bad reputation Alexandra had been given, her presence would have deprived them of credibility and embarrassed them in their fight against Bolshevism. Above all, it would have alienated the governments of Paris and London, which were supporting them. As Robert Wilton, the anti-Bolshevik correspondent of *The Times*, put it: 'Commandant Lasies, even if the Tsar and the Imperial Family are alive, it is necessary to say that they are dead.' On Kolchak's orders, therefore, Sokolov had to prove that the family were all dead. And doubtless he believed that he had proved it. But his zeal made him odious in the eyes of some of the Romanovs, who subsequently would not even meet him.

Since those days, the murder of the whole family having been accepted as established fact and the Anastasia affair dismissed as imposture, while other Romanovs had been murdered in the meantime, dynastic and anti-Bolshevik interests could not, without going back on their claims, allow the traditional version to be questioned. Nevertheless, Marina Grey, General Denikin's daughter, has, after all, yielded to doubt.[10]

The paradoxical feature of all this is that, for half a century, the Bolsheviks' interest and that of the dynasty have gone hand in hand. In the USSR even today the 'revelations' amount to reproducing the White version, with just a few additions to make the story spicier. What is the explanation for this?

The first reason, already mentioned, is that it was no longer necessary to dupe the Germans about the family's fate once the Hohenzollerns had fallen, in November 1918, and a few years later it ceased to be necessary to pretend to the British either, since, after the Genoa conference, Soviet Russia was accepted into the club of the Great Powers. The extent to which the family had been exterminated no longer mattered.

In Russia, where the general public knew nothing of the statements by Zinoviev, Chicherin and Litvinov, it was more convenient to say that all the Romanovs had been killed than to reveal (if this was indeed the case) that a negotiation had led to the release of Alexandra and her daughters, whereas the Tsar had been executed. The Soviet regime had nothing to gain by revealing such a negotiation, whatever its outcome had been. In a country where, more than anywhere else, historiography is under surveillance, no professional historian has thought until now of looking into the validity of the information that supports the hypothesis that Alexandra and her daughters survived: for example, of examining closely the negotiations between

---

10. She thinks that the members of the family who were taken to Perm were all shot later.

Bolsheviks and the Germans between April and November 1918, the roles played by Radek, Chicherin, and the rest. We have only the German and Spanish sources to go on.

Today, when *glasnost* calls for more light, it seems that the Soviets have not changed their original position. Discussion goes on around the material data of the crime, whether the bodies were burnt or buried, and so on. In other words, the argument relates only to technical matters and details that were already more or less available in works published by the Whites. Yurovsky's confession bears a curious resemblance to Medvedev's testimony. What is involved here is the revival of religious ideas, the rehabilitation of the Tsarist regime, and so on. The name of Yurovsky has re-emerged; he was a Jew and did not appear in Bykov's account published in 1926. Now he reappears so as to feed, as if need there was for this, the current stream of anti-Semitism. In Nicholas's time, as we have seen, every terrorist was assumed to be Jewish. Ryabov's account published in *Rodina* is significant, in that what is implicit in it is that the murder of the Russian Tsar was organized by two Jews, Sverdlov and Yurovsky; Lenin's name does not appear.

It is not these archives that will make it possible to check the hypothesis of survival, for they lie just as the institutions and authorities that secrete them lie. Who are we to believe – Medvedev and Yurovsky, or Dr Utkin and Maria Nikolaevna? Nor can we have confidence any longer in the analysis of objects, teeth and skulls, for we have long known that one expertise cancels out another.

Only a full confrontation of the Soviet archives with the relevant foreign archives could enable us to be more positive about what happened at Ekaterinburg than about Pope Joan. We can be certain that the British and the Danes could help in this matter, since they must possess information with a bearing on it, and that is also true of the Soviets. But what means are available to those who are interested in pursuing this story further? And where is the planet on which the story of 'what actually happened' will be told without hindrance, not by the courts or the state but by historians?

Whereas the deaths of Charles I and Louis XVI were great

historical events, that of Nicholas II provides an example of an item of 'news-in-brief', a non-event. Soviet historiography is silent about it. The Western historians who have mentioned it have done so either to adopt Sokolov's version or to question it, but not to integrate the event into the history of Russia.

The way that Lenin mentions it is significant: in parenthesis, and to recall that Russia was late on the scene when it came to executing monarchs, compared with Britain and France, who had done that long before. (Without real result, actually, since restoration followed.) What was important, Lenin explained, was to destroy the landlords and kulaks. The death of Nicholas II called for no explanation or justification. Lenin alluded to it as though to a report of something he had nothing to do with, and which was so unimportant that it was not worth spending any time on it.

That had not prevented him from negotiating with the Germans about the fate of the Tsaritsa and her daughters. But these secret talks – condemned only a short time before – have remained carefully hidden under the cloak of a theory of history that discounts individuals and knows only classes and modes of production. That is how such a negotiation and such an execution could become non-events, vanishing from history.

The paradox is that the same thing happened in the opposite camp. The fate of the imperial family was not made a political issue because the Whites knew that the unpopularity of Tsarism would harm their cause. It was enough to publicize the crimes of the Reds and just add thereto the murder of the Romanovs. Even the courts of Europe stood back from this event, intervening only for humanitarian reasons, never (the Hohenzollerns excepted) out of political considerations.

A non-event for the Reds, a non-event for the anti-Bolsheviks, the crime of Ekaterinburg found its place only in the 'news in brief' column.

True, between the idea that people formed of the Russian Revolution, on the basis of the sacred texts, and the reality of that revolution there is a gap no less wide than that which separates the reality of the Middle Ages from the Lives of the Saints. Contrasted with the Bolshevik leaders' talk of 'the triumphal march of Soviet power', the picture actually presented by the new post-October society is like a reversed image of that notion. Each town seeks to control its own railway, and the railway workers form themselves into an independent power. Prisoners of war rule the roost; the Communist Party is nowhere to be seen; only the Cheka and the soviet remind us that the revolution has put Lenin in power. In the White-occupied areas, to be sure, it is not clear who wields power, the military or the civilian authorities. Four different instances are involved in the judicial inquiry into the end of the Romanovs – or, rather, they either ignore or fight each other. In this context the end of the Romanovs not only figures in the 'news in brief' column, it even looks as though it belongs there.

Hardly have they learned, in Ekaterinburg, of the execution of the Tsar than the White officers who have just entered the town proceed to ransack the Ipatiev house, to which the townspeople also come, to look for souvenirs and touch objects that belonged to the victims. As in an ordinary crime case, the different instances of the state quarrel over the materials of the investigation, one magistrate after another being removed from it. Each instance has its own methods, selects its witnesses and puts forward its conclusions – all different. The most important witnesses disappear, executed or assassinated, always in mysterious circumstances. We cannot make out who is dead and who has escaped death. Nor can we see who bears responsibility for the executions: the Cheka, the soviet, the army's security service? We cannot understand why the testimony of those who give the orders contradicts that of the participants, something that nobody comments on – any more than on the disappearance of Goloshchekin and Beloborodov. In short, seventy years after, just as for the assassination of President Kennedy, the plentifulness of testimonies means that we cannot be sure what crime we are actually faced with.

Moreover, responsibility for the crime has never been claimed. The official communiqué does, of course, justify it, but no one actually takes responsibility. *You* have done well to do as you have done, says Moscow, and we do not really know who it was who gave the order to act – the local soviet, the Cheka, the central government, all three? For the Bolsheviks the blame lies with the Left SRs and with the Czechoslovaks and imperialists. For the Whites the blame lies with the Jews and the Austro-Hungarian prisoners of war. Or else, for both Whites and Reds, the crime was committed by Latvians. In any case, not by true Russians: they, the workers of Sysert, took no part in it, except for one of them, who denies it.

Today nothing has changed. The name of Yurovsky, who was a Jew, is brought forward again: and, as Beloborodov was not Jewish, the newspapers are quoting Nicholas who, in 1918, is supposed to have expressed surprise, because he thought the heads of all the organs of Soviet power were 'Bolshevik Jews'. Clearly, for the Whites as for those who today sing the memory of the murdered Tsar, he could not have been killed by true Russians – forgetting that it was true Russians who nicknamed him Nicholas 'the Bloody', and that the 300,000 workers who in January 1917 commemorated Bloody Sunday were true Russians. Today a speaker in the broadcast programme *The Fifth Wheel* dates the beginning of the terror directed against *the Russians* from the death of the Tsar, forgetting that the Red terror had begun long before that, against both Russians and non-Russians, and the White terror, too, for that matter.

But the formulation that associates the Bolshevik Revolution with an anti-Russian movement is interesting. In the time of *glasnost* it would be paradoxical if falsification of history were to advance faster than research into what actually happened. On the pretext that for a long time history has been manipulated in the USSR, any assertion at all, however absurd, could meet with welcome as a bird of freedom, assured of love and devotion. That situation has to be challenged.

Bibliography

I list here the books, articles and other documents that I used in writing this book. Those that I consider most essential for the subject are preceded by an asterisk. The figures in brackets indicate the chapter(s) to which particular works are relevant.

Soviet Archives

Ts GAORSSR: ff. 3, 130, 348, 406, 1235, 1244.
Ts GAORSSMo: ff. 1, 3, 66, 914.
Ts GAORSSLo: ff. 6384, 7384.
Ts GIAL: f. 1278.

See also the documents included in the works listed under Alexis de Durazzo, Ross, Seco Serrano and Sokolov, and the principal collections mentioned in Ferro, 1972, bibliography.

Actes du colloque 1905, edited by F. X. Coquin and C. Gervais-Francelle, Paris, 1984. (1, 2)

Akashi, M., *Rakka Ryusūi* (secret report on relations between the Japanese and the Russian revolutionaries), Helsinki, 1988.

Alexander Mikhailovich, *Once a Grand Duke*, London, 1932.

Alexandrov, V., *The End of the Romanovs*, London, 1966. (4)

Alexinsky, G., *Maxime Gorki*, Paris, 1950; 284f. (2)

Alexis de Durazzo, Prince d'Anjou, *Moi, Alexis, arrière-petit-fils du tsar*, Paris, 1982. (4)

Anweiler, O., *The Soviets, 1905–1921*, New York, 1975. (2, 3)

Avrekh, A., *Stolypin i tretia Duma* ('Stolypin and the Third Duma'), Moscow, 1968. (2)

Baynac, J., *Les Socialistes-revolutionnaires*, Paris, 1979. (1, 2)

Berard, V., *L'Empire russe et le tsarisme*, Paris, 1906. (1, 2)

Bogdanovich, A. V., *Journal*, Paris, 1926. (1)

Botkin, G., *The Real Romanovs*, London, 1932. (4)

Buchanan, Sir George, *My Mission to Russia*, London, 1923. (3, 4)

Buchanan, Meriel, *The Dissolution of an Empire*, London, 1932. (2, 3)

Buxhoeveden, Baroness, *Left Behind: Fourteen Months in Siberia during the Revolution*, London, 1928. (4)

Bykov, P. M., *The Last Days of Tsardom*, London, 1937 (Eng. trans. of *Poslednie dni Romanovykh*, Sverdlovsk, 1926). (4)

Bykov, P. M. and Niporsky, N., *Rabochaia revoliutsia na Urale* ('The Workers' Revolution in the Urals'), Ekaterinburg, 1921. (4)

Cantacuzène-Speransky, Princess, *Revolutionary Days*, London, 1920. (2, 3)

Chamberlin, W. H., *The Russian Revolution*, 2 vols, London, 1935. (3, 4)

Charques, R., *The Twilight of Imperial Russia*, Oxford, 1958. (1, 2)

Cherniavsky, M., *Prologue to Revolution*, Englewood Cliffs, NJ, 1967. (3)

—, *Tsar and People*, New Haven, Conn., 1961. (1)

Chmielewski, E., 'Stolypin's last crisis', *California Slavic Studies* (Berkeley, Calif.), III, 1964, pp. 95–127. (2)

Chute du régime tsariste, La, Paris, 1927.

Cinéma russe avant la révolution, Le, Paris, 1983.

Coquin, F. X., *La Révolution russe manquée, 1905*, Brussels, 1988. (1, 2)

Debo, R. K., *Revolution and Survival: The Foreign Policy of Soviet Russia, 1917–1918*, Liverpool, 1979. (4)

Dehn, L., *The Real Tsarina*, London, 1922. (2)

Dictionary of the Russian Revolution, edited by George Jackson and Robert Devlin, New York, 1989. (2, 3)

Dieterichs, G., *Ubiistvo Tsarskoy Sem'i* ('The Murder of the Imperial Family'), 2 vols, Vladivostok, 1922. (4)

Essad Bey, *Nicholas II*, London, 1936. (4)

Fedyshyn, O. S., *Germany's Drive to the East and the Ukrainian Revolution*, New Brunswick, NJ, 1971. (3, 4)

Ferro, M., *The Russian Revolution of February 1917*, London, 1972. (3)

—, *October 1917*, London, 1980. (3)

Field, D., *Rebels in the Name of the Tsar*, Boston, Mass, 1976. (2)

Freeze, G. L., 'A national liberation movement and the shift in Russian liberalism, 1901–3', *Slavic Review*, March 1969. (1)

Gajda, R., *Mje Pameti*, Prague, 1924. (4)

Geyer, D., *Der Russische Imperialismus*, Göttingen, 1977. (1, 2)

Gilliard, P., *Thirteen Years at the Russian Court*, London, 1921. (4)

Girault, R., *Emprunts russes et investissements français en Russie*, Paris, 1973. (1)

—, *Diplomatie européene et impérialisme 1871–1914*, Paris, 1979. (2)

Grey, Marina, *Enquête sur le massacre des Romanov*, Paris, 1987.

Grunwald, C. de, *Le Tsar Nicolas II*, Paris, 1965. (1, 2)

Haimson, L., 'Changement démographique et grèves ouvrières à St Petersbourg 1905–1914', *Annales*, 4, 1985.

Harcave, S., *The Years of the Golden Cockerel*, London, 1970.

Hasegawa, T., *The February Revolution*, Seattle, Wash., 1981.

Histoire de la littérature russe, edited by E. Etkin, G. Nivat *et al.*, Paris, 1987–8.

Hosking, G. A., *The Russian Constitutional Experiment: Government and Duma 1907–1914*, Cambridge, 1973. (3)

Ingerflom, S. C., 'Les Socialistes russes face aux pogroms, 1881–1883', *Annales*, 3, 1982. (1)

Ioffe, G. Z., *Velikii Oktyabr i epilog tsarisma* ('Great October and the Epilogue of Tsarism'), Moscow, 1987. (4)

Jones, David R., 'Nicholas II and the Supreme Command: an investigation of motives', *Sbornik*, No. 11, 1985.

Katkov, G., *Russia 1917: The February Revolution*, Oxford, 1967. (3)

Kerensky, A., '*The road to the tragedy*', appendix in P. Bulygin, *The Murder of the Romanovs*, London, 1935. (3, 4)

Koefoed, C. A., *My Share in the Stolypin Agrarian Reforms*, Odense, 1985. (2)

Kokovtsov, Count V. N., 'La vérité sur la tragédie d'Ekaterinbourg', *Revue des deux mondes*, October 1929. (4)

—, *Out of My Past*, Stanford, Calif., 1935.

Kondratieva, T., *Bolcheviks et Jacobins*, Paris, 1989. (1)

Kyril Vladimirovich, *My Life in Russia's Service*, London, 1939. (3, 4)

Laran, M. and Van Regenmorten, J. L., *Russie–URSS 1870–1984*, Paris, 1986. (1, 2)

Lasies, J., *La Tragédie sibérienne*, Paris, 1920. (4)

Legras, J., *Mémoires de Russie*, Paris, 1921. (3)

Lenin, V. I., *Collected Works*, 4th edn, English version, London, 1960–80.

Leontovitsch, V., *Geschichte d. Liberalismus in Russland*, Frankfurt, 1957. (1)

Leroux, G., *L'Agonie de la Russie blanche 1905*, Paris, 1928, reprinted 1978. (2)

*Leroy-Beaulieu, A., *The Empire of the Tsars and the Russians*, 3 vols, New York, 1893–6. (1)

Leyda, J., *Kino: A History of the Russian and Soviet Film*, 3rd edn, London, 1983.

Lockhart, R. Bruce, *Memoirs of a British Agent*, London, 1932.

Löwe, H. D., *Antisemitismus und reaktionäre Utopie. Russisches Konservatismus in Kampf gegen den Wandel von Staat und Gesellschaft 1890–1917*, Hamburg, 1978.

Maklakov, V. A., *The First State Duma*, Bloomington, Ind., 1964. (2)

Manning, R. T., *The Crisis of the Old Order in Russia*, New York, 1982. (2)

Marcadé, J. C., *Le Futurisme russe*, Paris, 1989. (2)

Markov, V., *Russian Futurism: A History*, Berkeley, Calif., 1968. (2)

Massie, R. K., *Nicholas and Alexandra*, London, Gollancz, 1968. (1, 2)

Max, Prince of Baden, *Erinnerungen und Dokumente*, Stuttgart, 1928. (4)

Melgunov, S., *Sud'ba Imperatora Nikolaya II posle otretseniya* ('The Fate of Nicholas II after His Abdication'), Paris, 1951. (4)

Milyukov, A. N., 'Dnevnik: peregovory s nemtsamy v 1918' ('Diary: negotiations with the Germans in 1918'), *Novyi Zhurnal*, 1961. (4)

Monas, S., *The Third Section*, Cambridge, Mass., 1961. (1)

Musée pittoresque du voyage du tsar, Paris, 1896. (1)

Nicholas II, *Archives secrètes*, Paris, 1928. (3)

—, *Dnevnik 1890–1906*, Berlin, 1923. (1, 2)

—, *Journal*, Paris, 1925. (1, 3)

—, *Journal intime 1914–1918*, Paris, 1934. (3)

—, *Letters of Tsar Nicholas and Empress Maria*, London, 1937. (2)

Nicholas Mikhailovich, *La Fin du tsarisme*, Paris, 1968. (3)

Nivat, G., 'Aspects religieux de l'athée russe', *Cahiers du monde russe et soviétique*, 1988, pp. 415–27.

Noguez, D., *Lenine dada*, Paris, 1989.

*Oldenburg, S. S., *Last Tsar*, 4 vols, Gulf Breeze, Fla, 1975–8. (1, 2)

Oukhtomsky, Prince, *Voyage en Orient de SAI le Tsarevitch*, Paris, 1895. (1)

Palat, M. K., 'Police socialism in Tsarist Russia, 1900–1915', *Studies in History* (New Delhi), II, 2, 1986, pp. 71–136.

Paléologue, M., *An Ambassador's Memoirs*, 3 vols, London, 1923–5.

Pamyatnaya Kniga na 1900, St Petersburg, 1900. (2)

Pares, Bernard, *The Fall of the Russian Monarchy*, London, 1939. (1, 2)

Pearson, R., *The Russian Moderates and the Crisis of Tsarism, 1914–1917*, London, 1977. (3)

Pélissier, J., *La Tragédie ukrainienne*, Paris, 1919, reprinted 1988.

Philippot, R., *La Modernisation inachevée 1855–1900*, Paris, 1974. (1)

Pipes, R., *P. Struve, Liberal on the Left*, Cambridge, Mass., 1970. (1)

Proyart, J. de, 'Le haut procureur du saint-synode Constantin Pobedonoscev et le "coup d'état" du 29 avril 1881', *Cahiers du monde russe et soviétique*, July–Sept. 1962. (1)

Prozhektor, 4, 1929. (4)

Radziwill, Princess Catherine, 'Was Tsar's family really slain?', *San Francisco Sunday Chronicle*, 11 July 1920. (4)

Renouvin, P., *La Question d'Extrême-Orient*, Paris, 1940. (1, 2)

Rodzianko, M. V., *The Reign of Rasputin*, London, 1927. (2, 3)

*Rogger, H., *Russia in the Age of Modernization and Revolution 1881–1917*, London 1983. (1, 2)

—, 'The formation of the Russian Right, 1900–1906', *California Slavic Studies* (Berkeley, Calif.), III, 1964, pp. 66–95. (2)

*Ross, Nikolai (ed.), *Gibel' Tsarskoy Sem'i* ('The Destruction of the Imperial Family'), Frankfurt, 1987. (4)

Ryabov, G. and Ioffe, G. Z., 'Prinuzhdeny vas rasstrelyat' ('We have to shoot you'), *Rodina*, 4 and 5, 1989. (4)

Sazonov, S., *Fateful Years, 1909–1916*, London, 1928. (3)

Seco Serrano, Carlos, *Viñetas históricas*, Madrid, 1983. (4)

Serebnikov, A., *Ubiistvo Stolypina* ('The Assassination of Stolypin'), New York, 1986. (2)

Seton-Watson, Hugh, *The Decline of Imperial Russia*, London, 1952. (1, 2)

Shavelsky, Father G., *Vospominaniya*, 2 vols, New York, 1954.

Silent Witnesses: Russian Films 1908–1919, London, 1989.

Sokolov, N., *Enquête judiciaire sur l'assassinat de la famille impériale russe*, Paris, 1924. (4)

Solovev, Yu. B., *Samoderzhavie i dvoryanstvo v kontse 19 veka* ('The Autocracy and the Nobility at the End of the Nineteenth Century'), Moscow, 1973. (1)

Speranski, V., *La Maison à destination speciale*, Paris, 1929. (4)

Spiridovich, A., *Les Dernières Années de la cour à Tsarskoïe Selo*, 2 vols, Paris, 1928. (1, 2)

'Staryi Professor', *Imperator Nikolai II i ego tsarstvovanie, 1894–1917* ('The Emperor Nicholas II and His Reign'), Nice, 1928. (1, 2)

Stavrou, T. G. (ed.), *Russia under the Last Tsar*, Minneapolis, Minn., 1969 (especially the contributions by T. Riha and D. W. Treadgold). (2)

Stone, Norman, *The Eastern Front, 1914–1917*, Oxford, 1975. (3)

*Summers, Anthony and Mangold, Tom, *The File on the Tsar*, London, 1976.

Surgević, Ilya, *Detsvo Imperatora Nikolaia II* ('The Childhood of the Emperor Nicholas II'), Paris, 1953. (1)

Trotsky, L. D., *1905*, London, 1972. (2)

Vernadsky, G. (ed.), *A Source Book for Russian History from Early Times to 1917*, 3 vols, New Haven, Conn., 1972. (1, 2)

Villemarest, P. E. de, *Le Mystérieux survivant d'octobre*, Geneva, 1984.

Vyrubova, A., *Souvenirs de ma vie*, Paris, 1927. (3, 4)

—, (?), *Journal secret, 1909–1917*, Paris, 1928. (3)

Walicki, A., *A History of Russian Thought, from the Time of Enlightenment to Marxism*, Stanford, Calif., 1979. (1)

Waters, W. H. N., *Secret and Confidential*, London, 1926. (1)

Wildman, A. K., *The End of the Russian Imperial Army*, 2 vols, Princeton, NJ, 1979. (2, 3)

Wilhelm II, *Correspondance entre Guillaume II et Nicolas II, 1894–1914*, Paris, 1929. (1, 2)

Wilton, R., *The Last Days of the Romanovs*, London, 1920. (3, 4)

Witte, S., *Memoirs*, London, 1921. (1, 2)

Wortman, Richard, 'Moscow and Petersburg: the problem of political centre in Tsarist Russia', in S. Wilentz (ed.), *Rites of Power since the Middle Ages*, Philadelphia, Pa., 1985, pp. 245–71. (1)

Youssoupoff, Prince, *La Fin de Raspoutine*, Paris, 1927. (3)

Zenkovsky, A., *Stolypin, Russia's Last Great Reformer*, Princeton, NJ, 1986. (2)

Index